ASBESTOS

Directory of Unpublished Studies

Second Edition

T0347969

Sponsored by the Commission of the European Communities and the Government of Canada in cooperation with the Asbestos Institute, Montreal.

ASBESTOS
Directory of Unpublished Studies

Second Edition

Editor

S. AMADUCCI

Lawyer, European Affairs, Brussels, Belgium

Taylor & Francis
Taylor & Francis Group

LONDON AND NEW YORK

Published by Taylor & Francis
2 Park Square, Milton Park, Abingdon, Oxon OX14 4RN
52 Vanderbilt Avenue, New York, NY 10017

First issued in paperback 2020

Taylor & Francis is an imprint of the Taylor & Francis Group, an informa business

British Library Cataloguing in Publication Data

Asbestos: directory of unpublished studies.
  —2nd ed.
  1. Asbestos—Research—Bibliography
  I. Amaducci, S.
  016.5536'72   Z7914.A7

ISBN 1-85166-073-9

Library of Congress CIP data applied for

Publication arrangements by Commission of the European Communities, Directorate-
General Telecommunications, Information Industries and Innovation, Luxembourg

EUR 7810 (second edition)

LEGAL NOTICE
Neither the Commission of the European Communities nor any person acting on behalf of the
Commission is responsible for the use which might be made of the following information.

ISBN 13: 978-0-367-58030-8 (pbk)
ISBN 13: 978-1-85166-073-5 (hbk)

CONTENTS

TABLE DES MATIÈRES

PREFACE

It was in 1981 that the joint sub-committee 'minerals and metals', set up in the framework of EEC–Canadian cooperation, decided to launch a publication which would provide an overall view of the progress achieved at international level in the field of research into asbestos. The publication, which appeared in 1982 under the title *Asbestos: Directory of Unpublished Studies* (publisher: D. Reidel, Dordrecht, The Netherlands), reflected both research still in progress and that completed but not yet published. To complement this, a directory of research and documentation centres particularly concerned with the problems of asbestos was published in the same year (title: *Asbestos: Directory of Research and Documentation Centres*, same publisher). The publication dealing with unpublished studies was to be updated for the first time the following year, with restricted circulation, together with a list of bodies which provide funds for research into asbestos.

In 1985 the Commission of the European Communities, which had thus far financed the publications, decided jointly with the Canadian Government that the directory of unpublished studies should be updated again. On this occasion the parties concerned, in cooperation with the Asbestos Institute, bore the costs involved in the new publication, which is up-to-date as at 31 October 1985. The main part of this Directory is in English. Only the introductory section (Preface and Stucture of the Directory) also appears in French.

As regards the contents of the Directory, it had emerged in previous editions that coverage was confined chiefly to health aspects. An attempt has therefore been made to extend the scope of this publication to cover technology and asbestos substitutes. It is the result of a survey comprising over 2000 research centres or individual researchers, who will afterwards be regularly consulted, as will the specialists who have been identified in the meantime. Future updates will probably be on an annual basis. As in the past, the two main objectives of this publication are to promote international cooperation and to prevent duplication. It is only in the long term that it will be possible to assess its achievements.

PRÉFACE

C'est en 1981 que le sous-comité mixte 'minéraux et métaux' établi dans le cadre de la coopération CEE–Canada décidait de lancer une publication destinée à rendre compte des progrès réalisés à l'échelle internationale dans le domaine de la recherche sur l'amiante. Cette publication, parue en 1982 sous le titre *Asbestos: Directory of Unpublished Studies* (D. Reidel éditeur, Dordrecht, Pays-Bas), reflétait les études en cours et aussi les recherches ayant abouti mais dont la publication n'avait pas encore été effectuée. Un ouvrage parallèle voyait le jour la même année sur les centres de recherche et de documentation ayant un intérêt particulier pour les questions relatives à l'amiante (titre: *Asbestos: Directory of Research and Documentation Centres*, même éditeur). La publication touchant aux études non publiées devait faire l'objet, l'année suivante, d'une première mise à jour à diffusion restreinte, qui était complétée par une liste des organismes pourvoyeurs de fonds pour la recherche sur l'amiante.

En 1985 la Commission des Communautés européennes, qui commanditait les publications parues jusque là, décidait en commun avec le Gouvernement Canadien une nouvelle mise à jour du répertoire sur les études non publiées. A cette occasion, les parties intéressées, en coopération avec l'Institut de l'Amiante, prenaient en charge les frais inhérents à cette nouvelle publication, dont les données reflètent la situation au 31 octobre 1985. La partie essentielle de ce répertoire a été rédigée en anglais. Seule la partie introductive (préface et structure du répertoire) comporte une version française.

Sur le plan du contenu de cette publication, il est apparu que les éditions antérieures ne rendaient compte, pour l'essentiel, que des études effectuées dans le domaine de la santé. C'est pourquoi le présent ouvrage s'efforce d'élargir son champ également aux domaines de la technologie et des substituts de l'amiante. Il est le résultat d'une enquête menée auprès de plus de 2000 centres de recherche ou chercheurs individuels, qui seront par la suite régulièrement consultés, de même que les spécialistes qui auront été identifiés entre-temps. Les futures mises à jour seront vraisemblablement assurées à une fréquence annuelle. Comme par le passé, les deux objectifs principaux poursuivis par cette publication sont de promouvoir la coopération internationale et d'éviter les doubles emplois. Ce n'est que sur une longue période que pourront se mesurer les objectifs atteints.

1

STRUCTURE OF THE DIRECTORY

Each page of Section 1 of this Directory deals with a single study. Each study is classified in alphabetical order of firstly the country and then the research centre. Whenever a particular research centre is carrying out more than one study, a progressive number (1, 2, 3, etc.) is added to its name.

For each study, the following information - if available - is provided:
- title
- research protocol (summary)
- stages of development (start, end, publication: actual dates or estimates)
- person(s) responsible
- sponsor(s)
- cooperation
- publication(s) on the project
- codes (each corresponding to a specific subject matter)

To be able to include as many studies as possible in the space available, the number of publications relating to the project concerned has been restricted to three. In addition, each study has been coded by the researchers themselves (apart from cases where information was far too insufficient) according to the tables on pages 5 & 6, which classify the studies by subject.

Section 2 contains a classification of the studies by subject on the basis of these tables. Each study is designated by the name of the relevant Centre, followed, where appropriate, by the above-mentioned progressive number and classified in alphabetical order of the country.

STRUCTURE DU REPERTOIRE

Chaque page du présent Répertoire, Section 1, correspond à une seule étude.
Chaque étude est classée par ordre alphabétique du pays et du centre de
recherche concernés. Un nombre progressif (1, 2, 3, etc.) associé au nom d'un
centre indique que ce dernier mène plus d'une étude.

Chaque étude est caractérisée - autant que possible - par les informations
suivantes:

- titre
- protocole (résumé)
- étapes de l'étude (début, fin, publication: dates réelles ou estimées)
- personne(s) responsable(s)
- soutien financier
- coopération
- publication(s) relative(s) à l'étude
- codes (désignant les matières couvertes).

Pour des raisons d'espace disponible, le nombre de publications mentionnées
dans la rubrique correspondante a été limité à trois (les plus récentes). On
notera enfin que chaque étude a été codée sous la responsabilité du centre
concerné (sauf dans les cas où les indications fournies étaient manifestement
insuffisantes) selon les tableaux de classification figurant aux pages 5 et
6, qui servent à grouper par matière les études recensées.

C'est précisément sur la base de ces tableaux que la Section 2 de ce Répertoire opère un regroupement des études selon les matières traitées. Chaque
étude y est représentée par le nom du centre de recherche concerné, suivi le
cas échéant du nombre progressif mentionné supra et classé par ordre alphabétique du pays dont le centre relève.

CLASSIFICATION SCHEME

A. DISCIPLINES

0. Multi-disciplinary
1. Geology
2. Mining
3. Chemistry
4. Physics
5. Mineralogy
6. "Mineralurgy" (valorization of minerals)
   a) recovery
   b) classification
7. Metrology
8. Mechanics
9. Biology
   a) experimental (in vitro/in vivo)
   b) bio-medicine
10. Epidemiology
11. Industrial medicine
12. Environment
13. Economy
14. Legislation
15. Standardization
16. Other

B. TECHNIQUES

0. Multi-techniques
1. Prospecting
2. Extraction
3. Sampling and classification
4. Processing
5. Dust control
6. Medical control
7. Animal experiments
8. Handling
   a) transport
   b) packaging/labelling
9. Quality control
10. Waste management
    a) disposal
    b) valorization
11. Recycling
    a) re-use
12. Energy consumption
13. Marketing
14. Other

C. RAW MATERIALS

0. Any fibres
1. Chrysotile
2. Amphibole
3. Modified fibres
4. Artificial fibres
5. Other fibres
   a) natural
   b) artificial
6. Asbestos tailings
7. Other materials

D. INDUSTRIAL APPLICATIONS

0. Any applications
1. Asbestos cement
   a) cement sheets
   b) cement pipes
   c) other
2. Asbestos textiles
   a) yarn and cords
   b) woven goods
   c) other
3. Insulation materials
   a) fire-proof
   b) heat
   c) acoustical
   d) electrical
4. Friction materials
   a) brakes
   b) clutches
   c) bearings
   d) other
5. Sealing materials
   a) joints
   b) caulking
   c) other
6. Fillers
   a) plastics
   b) paintings
   c) joint cement
7. Paper products
   a) felts (flooring, roofing)
   b) beater-add gaskets
   c) pipeline wrap
   d) millboard
   e) commercial paper
   f) specialty paper
   g) other

8. Asphalt mix
   a) floor
   b) roof
   c) road
9. Filters and diaphragms
10. Refractory products
11. Gaskets and packings
    a) gaskets
    b) packings
12. Floor tiles/plastics
13. Other

E. SUBSTITUTES

0. Any substitutes
1. Aluminium sheet
2. Aramid fibres
3. Beater-saturation materials
4. Carbon
5. Ceramics
6. Clays
7. Concrete
8. Cork composition
9. Ductile iron

10. Fusion bonded epoxy coatings
11. Glass fibres
12. Graphite
13. Graphite TFE composite
14. Membrane single-ply
15. Metal
16. Mica
17. Mineralised wood
18. Nylon
19. Organic felts
20. Plastics
21. Platey talc
22. PMF
23. Polyethylene
24. Polymers
25. Polyurethane
26. PVC
27. Reinforced concrete
28. Rubber
29. Semimetallics
30. Staple glass
31. Steel fibres
32. TFE
33. Vegetable fibre sheet
34. Vitrified clays
35. Wollastonite
36. Wood shingle
37. Other

TABLEAU DE CLASSIFICATION DES ETUDES

A. DISCIPLINES

0. Multidisciplinaire
1. Géologie
2. Génie minier
3. Chimie
4. Physique
5. Minéralogie
6. "Minérallurgie" (valorisation des minerais)
   a) récupération
   b) classification
7. Métrologie
8. Mécanique
9. Biologie
   a) expérimentale (in vitro/
   b) bio-médecine   in vivo)
10. Epidémiologie
11. Médecine industrielle
12. Environnement
13. Economie
14. Législation
15. Normalisation
16. Autres

B. TECHNIQUES

0. Multitechnique
1. Prospection
2. Extraction
3. Triage et classification
4. Transformation
5. Contrôle des poussières
6. Contrôle médical
7. Expériences animales
8. Manutention
   a) transport
   b) emballage/étiquetage
9. Contrôle de qualité
10. Gestion des résidus
    a) élimination
    b) valorisation
11. Recyclage
    a) réutilisation
12. Consommation d'énergie
13. Mise sur le marché
14. Autres

C. MATIERES PREMIERES

0. Toutes fibres
1. Chrysotile
2. Amphibole
3. Fibres modifiées
4. Fibres synthétiques
5. Autres fibres
   a) naturelles
   b) artificielles
6. Résidus d'amiante
7. Autres matières premières

D. APPLICATIONS INDUSTRIELLES

0. Toutes applications
1. Ciment d'amiante
   a) plaques de ciment
   b) tuyaux de ciment
   c) autres
2. Textiles d'amiante
   a) fils et cordons
   b) produits tissés
   c) autres
3. Matériaux d'isolation
   a) ignifuge
   b) thermique
   c) acoustique
   d) électrique
4. Matériaux de frottement
   a) freins
   b) embrayages
   c) roulements
   d) autres
5. Produits d'étanchéité
   a) joints
   b) calfeutrage
   c) autres
6. Produits de charge et de remplissage
   a) plastiques
   b) peintures
   c) ciment de jointoiement
7. Produits papier
   a) feutres (dallage, toiture)
   b) "beater-add gaskets"
   c) couvre-joint de pipeline
   d) carton
   e) papier commercial
   f) papier à usages spéciaux
   g) autres

8. Mélange asphalte
   a) sols
   b) toitures
   c) routes
9. Filtres et diaphragmes
10. Produits réfractaires
11. Joints et garnitures
    a) joints
    b) garnitures
12. Carreaux/plastiques de carrelage
13. Autres

E. SUBSTITUTS

0. Tous substituts
1. Tôles d'aluminium
2. Fibres aramides
3. "Beater-saturation materials"
4. Carbone
5. Céramiques
6. Argiles
7. Béton
8. Agglomérés de liège
9. Fonte ductile
10. Revêtements en époxyde soudés par fusion
11. Fibres de verre
12. Graphite
13. Matériau composite de graphite TFE
14. Membrane à une couche
15. Métal
16. Mica
17. Bois minéralisé
18. Nylon
19. Feutres organiques
20. Plastiques
21. Talc en paillettes (Platey talc)
22. PMF
23. Polyéthylène
24. Polymères
25. Polyuréthane
26. PVC
27. Béton armé
28. Caoutchouc
29. Matériaux demi-métalliques
30. Verranne
31. Fibres d'acier
32. TFE
33. Panneaux en fibres végétales
34. Argiles vitrifiées
35. Wollastonite
36. Bardeaux en bois
37. Autres

7

Section 1

**DESCRIPTION OF STUDIES BY COUNTRY AND
IN ALPHABETICAL ORDER OF THE RELEVANT RESEARCH CENTRES**

Australia

ASBESTOS RESEARCH GROUP, QUEEN ELIZABETH II MEDICAL CENTRE (1).

Original language : English

Title : **MORTALITY IN THE CROCIDOLITE INDUSTRY OF WITTENOOM, WESTERN AUSTRALIA.**

Research protocol :

6,505 males and 411 females known to have been employed in the crocidolite industry at Wittenoom, Western Australia between 1943 and 1966 have been followed through various public records to December 1980 providing 95,264 person years of follow-up of males and 4,914 person years of follow-up for the females up to the date last known to be alive. Vital status and certified cause of death has been determined and standardized mortality ratios estimated for the coded cause of death.

Progress : start = 01/75   end = 12/85   publication = 03/86

Person(s) responsible : A.W. Musk, M.S.T. Hobbs, B.L. Armstrong.

Sponsor(s) : NH & MRC (Australia), CSR Pty. Ltd.

Cooperation : Sir Charles Gairdner Hospital and Public Health Department of Western Australia.

Publication(s) on the project :
Hobbs MST, Woodward SD, Murphy B, Musk AW and Elder JE. The incidence of pneumoconiosis, mesothelioma and other respiratory cancer in men engaged in mining and milling crocidolite in Western Australia in Biological Effects of Mineral Fibres, ed. Wagner JC. International Agency for Research on Cancer. Lyon 1980. pp. 615-25.

Codes : A.10 - C.5.a)

ASBESTOS RESEARCH GROUP, QUEEN ELIZABETH II MEDICAL CENTRE (2).

Original language : English

Title : **NATURAL HISTORY OF ASBESTOSIS IN FORMER CROCIDOLITE WORKERS AT WITTENOOM GORGE.**

Research protocol :

The course of pulmonary asbestosis and its determinants have been examined in 280 applicants for compensation among former workers of the crocidolite mine and mill at Wittenoom Gorge, Western Australia. Serial chest radiographs accrued over more than three decades were graded for parenchymal disease separately by two observers according to the 1980 ILO Classification of Radiographs for Pneumoconioses and without knowledge of exposure histories or compensation details. In 136 subjects whose median duration of exposure was 37 months, radiographic asbestosis appeared between 1 and 34 years after initial exposure and then progressed continuously. Total exposure to asbestos and time from first exposure to the appearance of definite radiographic asbestosis were significant determinants of the rate of progression of profusion of radiographic abnormality. Asbestosis should be considered to be an active disease aven three decades after exposure has ended.

Progress : start = 1983   end = 12/85   publication = 03/86

Person(s) responsible : A.W. Musk, M.S.T. Hobbs, B.K. Armstrong, W.O.C. Cookson

Sponsor(s) : NH & MRC (Australia), CSR Pty. Ltd.

Cooperation : Sir Charles Gairdner Hospital and Public Health Department of Western Australia.

Publication(s) on the project :

Codes : A.10 - C.5.a)

NATIONAL OCCUPATIONAL HEALTH & SAFETY COMMISSION.

Original language : English

Title : **AUSTRALIAN MESOTHELIOMA REGISTER.**

Research protocol :

Australia has a special problem with mesothelioma because of the extraction of crocidolite from about 1945 to 1966. The fibre type was thought particularly hazardous, and mesothelioma induly frequent. Standard of diagnosis was uncertain, a work history of asbestos exposure could not be obtained in many cases. Extensive information was gathered, for all known Australian cases, on pathology, occupational and environmental history, and type and number of fibres in lung tissue. The objects were to improve diagnosis, define occupational and other environmental associations, identify trends in occurrence, and further knowledge on natural history. Notifications have been received from pathologists, clinicians, cancer registrars, hospital records librarians and government occupational health authorities. Detailed data collection started in 1980 and finished on 31st December 1985, by which time over 900 cases had been registered. Ongoing notification in limited detail will follow for some years, to identify future trends. Analysis is proceeding.

Progress : start = 06/1977   end = 1987   publication = 1986-88

Person(s) responsible : Professor D.A. Ferguson.

Sponsor(s) : National Occupational Health & Safety Commission.

Cooperation : Hospitals, professional associations, cancer registers.

Publication(s) on the project : Rogers AJ. Ann. Occup Hyg, 1984; 28:1-12.

Codes : A.0 - B.6 - C.5 - D.0 - E.0

SYDNEY UNIVERSITY, DEPARTMENT OF MECHANICAL ENGINEERING.

Original language : English

Title : **FRACTURE OF CEMENTITIOUS COMPOSITES.**

Research protocol :

To develop asbestos-free cements using both wood and polymeric fibers. It is also intended to understand the fracture mechanics of cementitious matrices and their fiber composites.

Progress : start = 09/79   end =   publication =

Person(s) responsible : Associate Professor Y.W. Mai and Dr. B. Cotterell.

Sponsor(s) : James Hardie & Coy Pty Ltd (1979-1983); Australian Research Grants Committee (1985-).

Cooperation :

Publication(s) on the project :
1. Proc. Int. Conf. on Fracture Mechanics of Concrete: Lausanne, Switzerland, Oct. 1985.
2. J. Mater. Sci., 19(1984) 501-508.

Codes : A.8 - C.5 - D.1 - E.0 - E.2 - E.24.

Australia

WOODSREEF MINES LIMITED. (1)

Original language : English

Title : **USE OF WET PROCESS CHRYSOTILE IN ASBESTOS CEMENT.**

Research protocol :

> To evaluate the use of wet process chrysotile fibre in the production of asbestos cement products to determine (i) whether the fibre can be used on existing Hatschek machines (ii) whether products incorporating wet process fibre have any premium qualities and (iii) whether it can assist in improving dust control in AC manufacturers premises.

Progress : start = 5/82   end = 12/87   publication = 5/86

Person(s) responsible : Dr. P.S.B. Stewart

Sponsor(s) : Transpacific Resources Inc., Baie Verte Mines Inc., Woodsreef Mines Ltd.

Cooperation :

Publication(s) on the project :

> "Asbestos Cement Manufacture from Wet Process Chrysotile", First International Conference on Asbestos Cement, Cannes, May, 1986.

Codes : A.12 - B.4 - B.5 - B.13 - C.1 - C.3.

Australia

WOODSREEF MINES LIMITED. (2)

Original language : English

Title : **RECOVERY OF CHRYSOTILE BY WET PROCESSING.**

Research protocol :

To develop an economic process for recovery of asbestos cement grade chrysotile fibre from ore or tailings by means of a wet process which reduces atmospheric dust in processing to levels close to the normal limit of detection. A tailings which is virtually barren to be discarded and the product fibre to be in a form which can be readily handled without the production of dust.

Progress : start = 08/77   end = 12/87   publication = 5/86

Person(s) responsible : Dr. P.S.B. Stewart

Sponsor(s) : Transpacific Resources Inc., Baie Verte Mines Inc., Woodsreef Mines Ltd.

Cooperation :

Publication(s) on the project :

1. "Benefication of Chrysotile by Wet Processing – the Resource without the Hazard." To be published in the 13th Congress of the Council of Mining & Metallurgical Institutions, Singapore, May, 1986.
2. "Woodsreef Asbestos Tailings Retreatment by Wet Processing" Environmental Impact Statement, Woodsreef Mines Ltd. March, 1985. (unpublished).

Codes : A.6.a) – A.12 – B.4 – B.5 – B.8.b) – B.10 – C.1 – C.3 – C.6

Austria

INSTITUTE OF ENVIRONMENTAL HYGIENE, UNIVERSITY OF VIENNA (1).

Original language : English

Title : **MORTALITY, MORBIDITY, CHEST X-RAY AND LUNG FUNCTION IN ASBESTOS CEMENT WORKERS.**

Research protocol :

In a historical prospective cohort study asbestos cement workers showed increased mortality from lung cancer compared to cement workers ($p < 0.05$), smoking cement workers ($p < 0.05$) and the local population (n.s.). Cohorts have been enlarged and other data on mortality, morbidity, chest X-ray and lung function have beend collected as well as more detailed exposure and smoking histories. The evaluation of an intervention study will start in 1986. Some preliminaries and partial results have been published.

Progress : start = 1976 (1950)   end =   publication =

Person(s) responsible : M. Neuberger, M. Haider

Sponsor(s) : Several (Austrian Science Research Fund and other)

Cooperation : Health authorities, work inspection, workmen's compensation board, etc.

Publication(s) on the project :

1. Prevention of occupational cancer such as in the asbestos cement industry (Preliminary Zbl. Bakt. Hyg. IB 181, 81–86 (1985).
2. Österreichischer Ringversuch zur Röntgen-Frühdiagnose der Asbestose, ASP 20, 7, 159–162 (1985).

Codes : A.10 – A.11 – B.6 – C.1 – C.2 – C.7 – D.1

INSTITUTE OF ENVIRONMENTAL HYGIENE, UNIVERSITY OF VIENNA (2).

Original language : English

Title : **LOW LEVEL (ENVIRONMENTAL) ASBESTOS EXPOSURE AND HEALTH.**

Research protocol :

Population screening in an area with environmental asbestos exposure showed increased prevalence of pleural plaques compared to control regions, but so far no increased cancer mortality could be found (1). Lung function showed a steeper decline with age than predicted from different reference value (2). Samples of soil, air and lungs are analysed by light and electron microscopy, EDAX and LAMMA.

Progress : start = 1975   end =   publication =

Person(s) responsible : M. Neuberger, M. Haider.

Sponsor(s) : Several (Austrian Ministry of Health and Environmental Protection and other).

Cooperation : University of Soil Science Vienna, Technical University Vienna, University of Antwerpen.

Publication(s) on the project :

(1) Environmental asbestos exposure and cancer mortality. Arch Environ. Health 39, 4, 261-265 (1984).
(2) Lung function screening in a population with endemic pleural plaques related to environmental asbestos exposures. Wr. Klin. Wschr. 97, 6, 289-293 (1985).

Codes : A.10 - A.12 - B.0 - B.6 - C.2 - C.7 - D.12.

OSTERREICHISCHE STAUB (SILIKOSE)- BEKÄMPFUNGSSTELLE.

Original language : German

Title : **DETERMINATION OF AIRBORNE ASBESTOS FIBRE CONCENTRATIONS AND CLASSIFICATION OF WORKER EXPOSURE.**

Research protocol :

    Sampling.

    Analysing.

    Classification of worker exposure.

Progress : start = continuous   end =   publication =

Person(s) responsible : Dipl.-Ing. Eckhard Bigga.

Sponsor(s) : Allgemeine Unfallversicherungsanstalt (General Accident Insurance Organisation).

Cooperation : Federal Minister for Social Affairs.

Publication(s) on the project :

Codes : A.0 - B.3 - B.5 - C.0 - D.0

MICRO & TRACE ANALYSIS CENTRE (MITAC), UNIVERSITY OF ANTWERP.

Original language : Dutch

Title : **FIBER SURFACE CHARACTERIZATION OF INDUSTRIALLY TRANSFORMED ASBESTOS AND SUBSTITUTES BY LASER MICROPROBE MASS ANALYSIS (LAMMA).**

Research protocol :

Laser microprobe mass analysis with the commercial LAMMA-500 (transmission type) instrument, when utilized in laser desorption conditions, provides information which allows the identification of particular asbestos varieties, in various types of samples, including lung tissue. Also it has detection capability of adsorbed organic material and it is a valuable qualitative and semi-quantitative tool for the study of the surface of chemically modified and industrially transformed asbestos fibers and substitution products.

Progress : start = 01/10/82   end = 01/10/86   publication =

Person(s) responsible : Prof. Dr. F. Adams - J.K. De Waele.

Sponsor(s) : Commission of the European Communities.

Cooperation :

Publication(s) on the project : ca 20 till July 1985; summary of activities J.K. De Waele and F. Adams, "Applications of laser microprobe mass analysis for characterization of asbestos", Scanning Electron Microscopy, III (1985) 935-946.

Codes : A.0 - A.3 - A.5 - A.10 - A.12 - B.0 - C.0 - C.3 - C.4 - C.5 - D.0 - E.0.

SERVICE DE PNEUMOLOGIE, HOPITAL ERASME.

Original language : French

Title : **MINERALOGICAL ANALYSIS OF LUNG AND LAVAGE IN LUNG CANCER IN THE
GENERAL POPULATION.**

Research protocol :

Study of the clinical, functional, X-ray and mineralogical (optical and
electron microscopes) points of view of all patients undergoing surgical
removal of lung cancer in the hospital.
Comparison with lavage.

Progress : start = 1/10/85   end =   publication =

Person(s) responsible : P. De Vuyst.

Sponsor(s) : Fondation Roi Baudoin / Fondation Erasme / FRSM / United Fund.

Cooperation : Dr. J.C. Wagner, Penarth.

Publication(s) on the project :

Codes : A.0 - A.5 - A.7 - A.10 - B.6 - D.0

CASSIAR MINING CORPORATION.

Original language : English

Title : **ENHANCED RECOVERY OF ASBESTOS FIBRE USING A WET MILLING PROCESS.**

Research protocol :

To establish the methods and machinery necessary for the economic extraction and grading of asbestos fibre from mill tailings using a wet process. Emphasis to be placed on simplicity, low operating costs and the production of a high quality fibre capable of competing in international markets.

Progress : start = 01/05/85   end = 31/04/88   publication = 31/04/88

Person(s) responsible : G. Riley (Scientific Advisor - Canmet), S.M. Dyk (Mill Superintendent - Cassiar Mining Corporation).

Sponsor(s) : Joint sponsorship - Canmet and Cassiar Mining Corporation.

Cooperation : Ontario Research Foundation.

Publication(s) on the project :

Codes : A.6.a) - A.6.b) - B.2 - B.4 - B.5 - B.8 - B.8.a) - B.8.b) - B.9 - B.10.a) - B.12 - B.13 - C.6

21

CENTRE SPECIALISE EN TECHNOLOGIE MINERALE.

Original language : French

Title : **THE USE OF SLUDGE FROM SEWAGE WORKS IN THE SOWING OF ASBESTOS-BEARING ORE WASTE DUMPS.**

Research protocol :

The purpose of the project is to determine the conditions for the use of sludge in introducing and maintaining plant life on dumps of asbestos mine waste. The objectives are to exploit the potential of sewage sludge which has fertilising properties and to reintroduce vegetation to waste dumps. The protocol consists of the following stages:

1. The identification of the chemical properties of the sludge.
2. The determination of the physical and chemical properties of mining waste.
3. The selection, under glass, of the species of plant (grasses and legumes) which are able to grow in these special conditions i.e. in sludge and mining waste.
4. The proposing of the correctives (fertilisers) necessary for the adequate growth of certain species, after experimentation under glass.
5. The determination of the conditions for the stabilisation of the sludge on the receiving medium, in order to prevent erosion.
6. The examination of the quality of the run-off water, to make sure that toxic substances do not return to the watercourses.
7. The checking of the phytotoxicity of the receiving medium by analysing toxic and undesirable elements in the plants.
8. Verification that the vegetation continues to survive in these conditions.

Progress : start = 01/06/86   end =   publication =

Person(s) responsible : M. Réjean Nadeau, directeur.

Sponsor(s) : La Société québécoise d'assainissement des Eaux, Employment and Immigration Canada.

Cooperation : La Régie intermunicipale d'assainissement des Eaux de la Haute-Bécancour.

Publication(s) on the project :

Codes : A.0 - A.2 - A.3 - A.4 - A.5 - A.9 - A.12 - B.5 - B.10 - B.11 - C.6

DEPARTMENT OF COMMUNITY HEALTH SCIENCES,
FACULTY OF MEDICINE, THE UNIVERSITY OF CALGARY.

Original language : English

Title : **INDOOR AIR QUALITY OF HEALTH CARE FACILITIES.**

Research protocol :

> Hospital air is assumed to be cleaner than other industrial environments. This statement is true with respect to microbial contamination. However, increasing numbers of new chemicals are introduced to hospitals for diagnostic, therapeutic and bio-medical research purposes. Airborne and surface contamination of hospitals by chemicals may be studied in terms of indoor air quality in comparison between hospitals with traditional construction and others with energy efficient design. One of the chemical contaminants often argued among hospital maintenance managers is asbestos coatings for fire protection installed in older hospitals. Assessing of indoor air quality always indicates the presence of airborne asbestos fibres. Emission of asbestos fibres inside hospitals must be controlled without the extraordinary cost of increased ventilation.

> The research project intends to identify problem areas of indoor air quality in hospitals in terms of facility and building material. A list of critical contaminants contains non-viable particle and carbon dioxide as a priority group. Then the list goes to anti-neoplastic drugs and sterilant gas and asbestos. Research results will provide, as a part of integral IAD investigation, practical methods in managing asbestos in hospitals.

Progress : start = 01/10/85   end = 31/03/88   publication = 01/10/88

Person(s) responsible : Dr. Ken Yoshida, Associate Professor.

Sponsor(s) : Alberta Occupational Health & Safety Division, Health and Welfare Canada.

Cooperation : Foothills Provincial General Hospital, Calgary.

Publication(s) on the project : Yoshida, K.: Concentration of airborne non-viable particles and carbon dioxide in a teaching hospital, Qt. Jr. Can. Hosp. Eng. Soc. 5(3): 26-29, 1985.

Codes : A.12 - B.12 - C.1 - D.3.a)

Canada

DEPARTMENT OF PHYSICS, DALHOUSIE UNIVERSITY.

Original language : English.

Title : **MAGNETIC PROPERTIES OF ASBESTOS**.

Research protocol :

Using a vibrating sample magnetometer, a SQUID magnetometer and Mossbauer spectroscopy, the magnetic properties of UICC asbestos samples and lung tissues are being measured. The results are analysed in view of using magnetic measurements in the lungs of asbestos miners and millers to determine their dust load (magneto pneumography).

Progress : start = 01/04/83   end = 31/03/86   publication =

Person(s) responsible : Dr. G. Stroink.

Sponsor(s) : Natural Science and Engineering Research Council of Canada.

Cooperation :

Publication(s) on the project: G. Stroink, D. Hutt, D. Lim and R.A. Dunlap, "The magnetic properties of chrysotile asbestos", IEEE ~ Mag., MAG-21, 1985, "Biomagnetism: Applications and Theory", Eds. Weinberg, Stroink, Katila, Pergamon Press, N.Y., 1985; G. Stroink, "The magnetic properties of respirable asbestos in airborne dust and in lungs of asbestos workers.". Final report. Contract no. OST83-00068 of Nat. Research Council, Biol. Div., Ottawa, 91 pp., 1984.

Codes : A.4 - A.9.b) - C.0

DEPARTMENT OF SOIL SCIENCE, UNIVERSITY OF BRITISH COLUMBIA.

Original language : English

Title : **LINKING TRACE METALS WITH ASBESTOS FIBRES IN A RURAL ENVIRONMENT AFFECTED BY SERPENTINITIC SEDIMENTS.**

Research protocol :

A landslide has exposed serpentinitic bedrock in the headwaters of the Sumas River and in the process of weathering large quantities of chrysotile asbestos fibres are introduced into the stream system. During several flooding events asbestos-rich sediments have been deposited in agricultural fields creating a very poor media for plant growth. The material lacks major nutrients, contains high levels of trace metals and is hazardous to human health. Cr, Co and Ni associated with the asbestos fibres were used successfully as tracers of the asbestos fibre distribution in the environment which includes water, sediments, soils, plants and animals. The asbestos material has an adverse effect on stream quality causing Cr and Ni values to exceed drinking water standards and Cr, Ni and Co values in the sediments were significantly higher than the regional background levels over the entire 16 km river section examined in this study. The deposited sediments are not conductive to plant growth and trace metals leach at a slow rate into lower soil horizons. The trace metal levels in the few plants that have colonized inundated sites are higher than those in normal plants but they are low in comparison to levels found in native plants found on serpentinitic soils elsewhere. Experiments with earthworms showed that the animals were adversely affected by the asbestos-rich sediments, showing short term uptake of Ni and Mg and 100% long term mortality. Cattle grazing in the vicinity of the inundated sites was found to have higher Ni and Mn values in the blood than normal unaffected animals, and asbestos fibres were also detected in the blood samples. The trace metal values in the blood of affected animals rapidly returned to normal levels once the animals were removed from the contaminated site. Recommendations to reduce the health hazards are to revegetate the affected sites by the addition of organic matter and major nutrients and to maintain pH levels above 7.5 in the soil and stream.

Progress : start = 01/08/1983   end = 3/12/84   publication =

Person(s) responsible : H. Schreier.

Sponsor(s) : National Research Council, Associate Committee on Environmental Quality, Ottawa, Ont., Canada.

Cooperation :

Publication(s) on the project : 2 papers are in press, 2 will be forthcoming.
    Schreier, H., J.A. Shelford, and T.D. Nguyen (in press): Asbestos fibres and trace metals in the blood of cattle grazing in fields inundated by asbestos rich sediments. Envir. Res.
    Schreier, H. and H.I. Timmenga (in press): Earthworm response to asbestos rich sediments: in Soil Biology and Biochemistry, Dec. 1985, Issue.

Codes : A.0 - A.3 - A.9 - A.12 - B.0 - B.2 - B.3 - B.10 - C.1.

Canada

DEPARTEMENT DE CHIMIE, UNIVERSITE LAVAL.

Original language : French

Title : **CHEMICAL METHOD FOR MEASURING THE LEVEL OF CHRYSOTILE IN THE AIR.**

Research protocol :

Since optical microscopy presents certain weaknesses as a way of measuring the level of chrysotile in the air of asbestos mines, other methods (eg X-ray diffraction, infra-red spectroscopy and, in particular, electron microscopy) are being investigated. Research is also being carried out into finding a reliable chemical test. This has involved determining whether the level of magnesium released in an acid solution, provides an accurate indication of the quantity of chrysotile fibres in a given sample. Since chrysotile might not be the only source of magnesium in the sample, a method was found of distinguishing the chrysotile magnesium from that of other magnesium compounds, with the exception of two, namely, lizardite and forsterite. However, it is considered unlikely that these two compounds would be found in the air. The results of tests compared well with those carried out by optical microscopy, at both high and low levels of asbestos. The method now has to be tested out in the field.

Progress : start = 05/82   end = 03/85   publication =

Person(s) responsible : Prof. Claude Barbeau.

Sponsor(s) : Institut de recherche en santé et en sécurité du travail du Québec (IRSST).

Cooperation :

Publication(s) on the project :

Codes : A.3 - A.7 - B.3 - B.5 - C.1

Canada

DEPARTEMENT DE CHIMIE, UNIVERSITE DE SHERBROOKE.

Original language : French

Title : **STUDY INTO THE ADSORPTION CAPACITY OF CARCINOGENIC MOLECULES ON ASBESTOS OR OTHER PARTICLE MATERIALS.**

Research protocol :

A number of industrial processes using carbon at high temperatures release potentially very toxic substances, such as polyaromatic hydrocarbons, which dangerously pollute the atmosphere. The latter present well-known carcinogenic properties.

It is known that polyaromatic hydrocarbons are physically and chemically adsorbed by asbestos. The study of the adsorbtive properties of asbestos and other particle materials vis-à-vis toxic products, will lead to a better understanding of the causes of asbestosis and other lung diseases associated with certain workplaces. On the other hand, the adsorbtive nature of asbestos could make it suitable for use as a filter to eliminate these toxic products. An understanding of these aspects is essential for the elimination from industrial processes of health risks to workers, and at the same time could bring an end to the threat of closure faced by some of our industries.

Progress : start = 01/08/85   end = 31/07/86   publication =

Person(s) responsible : Hugues Ménard.

Sponsor(s) : Institut de recherche en santé et en sécurité du travail du Québec (IRSST).

Cooperation :

Publication(s) on the project :

Codes : A.3 - A.6 - A.10 - A.11 - A.12 - A.13 - D.9

FACULTE DE MEDECINE, DEPARTEMENT DE PNEUMOLOGIE, UNIVERSITE DE SHERBROOKE.

Original language : French

Title : **EVALUATION OF BIOLOGICAL EFFECTS ON THE LUNG OF FIBROUS SUBSTITUTES FOR ASBESTOS.**

Research protocol :

Study of fibrogenic potential, and effects on lung macrophages, of various fibrous substitutes for asbestos (naturally-occurring and man-made).

Progress : start = 01/02/86   end = 31/01/88   publication =

Person(s) responsible : Dr. Irma Lemaire.

Sponsor(s) :   Institut de l'amiante.

Cooperation :

Publication(s) on the project :

Codes : A.9.a) – C.1 – C.4 – C.5

FACULTE DES SCIENCES, DEPARTEMENT DE BIOCHIMIE, UNIVERSITE LAVAL.

Original language : French

Title : **EFFECTS OF FIBROUS MATERIALS ON NUCLEAR TOXICITY.**

Research protocol :

    Study of changes in poly (ADP - ribose) polymerase, and of implication of free radicals, after challenge with fibrous materials.

Progress : start = 01/04/86   end = 31/03/89   publication =

Person(s) responsible : Dr. Guy Poirier.

Sponsor(s) : Institut de l'Amiante.

Cooperation :

Publication(s) on the project :

Codes : A.9.b) - C.0

INSTITUT ARMAND-FRAPPIER, CENTRE DE RECHERCHE EN EPIDEMIOLOGIE. (1)

Original language : English

Title : **AN EXPOSURE-BASED CASE-CONTROL APPROACH TO STUDYING THE ASSOCIATIONS BETWEEN 14 SITES OF CANCER AND MANY OCCUPATIONAL EXPOSURES.**

Research protocol :

Since 1979 we have established a novel epidemiologic approach to investigate possible associations between cancer and environmental exposures in the workplace. It is a case-control type study whereby all patients in the Montreal area with newly diagnosed cancers of certain sites, among males aged 35-70 are interviewed by specially trained interviewers, as is a control series selected from the general population. Fourteen cancer sites have been included. Each subject is asked to enumerate the jobs and industries he has worked in and also to provide a detailed description of the tasks performed and materials manipulated in each job. Following the in-depth interview, the file containing descriptions of each job the man has held, is passed on to a team of specially trained chemists and hygienists. Their task is to list the chemicals to which the subject may have been exposed in each of his jobs. This task is accomplished with reference to a checklist which contains 275 potential occupational exposures. One of the potential exposures evaluated is asbestos. Several sources of information are used by the chemists, including the interview respondent himself, technical documentation and consultants. Each job environment is unique and is treated as such by our chemists. We have now interviewed nearly 4,000 cases, distributed unequally among 14 sites of cancer. For statistical analysis, each site series constitutes a case group, and can be compared with two control groups: the population controls and the other types of cancer. It is possible to analyse relative risks between any of the 14 sites of cancer and of the 275 chemical exposures on the checklist. In the next 18 months a special analysis will be undertaken to describe cancer risks associated with asbestos exposure across the gamut of occupations and industries.

Progress : start = 01/01/79   end = 31/12/87   publication =

Person(s) responsible : Jack Siemiatycki, Michel Gérin.

Sponsor(s) : Institut de recherche en santé et sécurité du travail du Québec, National Health Research and Development Program of Canada, National Cancer Institute of Canada.

Cooperation : 21 Montreal area hospitals.

Publication(s) on the project :
1. Siemiatycki J., Day N., Fabry J. and Cooper J. Discovering carcinogens in the occupational environment: a novel epidemiologic approach. J.N.C.I. 66:217-225, 1981.
(cont p. 193)

Codes : A.10 - C.0 - D.0

INSTITUT ARMAND-FRAPPIER, CENTRE DE RECHERCHE EN EPIDEMIOLOGIE. (2)

Original language : French

Title : **CANCER RISK DUE TO EXPOSURE TO ASBESTOS AND OTHER FIBROUS MATERIALS IN VARIOUS WORKPLACES: MONTREAL SURVEY.**

Research protocol :

Investigation into cancer risk due to exposure to any respirable material, including asbestos, in various trades excluding mining and milling.

Progress : start = 01/04/86   end = 31/03/88   publication =

Person(s) responsible : Dr. Jack Siemiatycki.

Sponsor(s) :   Institut de l'amiante.

Cooperation :

Publication(s) on the project :

Codes : A.10 - C.0 - D.1

31

Canada

LABORATOIRE DE BIOCHIMIE ET DE TOXICOLOGIE PULMONAIRES,
DEPARTEMENT DE BIOLOGIE, FACULTE DES SCIENCES, UNIVERSITE DE SHERBROOKE.

Original language : French

Title : **CYTOTOXICITY OF NATURAL AND MAN-MADE FIBRES.**

Research protocol :

The purpose of the project was to identify and compare the biological activity of mineral fibres (natural and man-made). On the basis of a material (fibres) clearly characterized from the physico-chemical point of view and available in sufficient quantity (20-25 mg/fibre), the aim was to analyse the biological activity of the various fibres chosen and to compare the effects obtained in a tissue culture with other methods of evaluation, such as tests for haemolysis, bioluminescence and DNA transformation (cancer).
The study covered the following areas:
- the determination of cytotoxicity connected with the various types of fibres in vitro;
- the identification of the fibres on which in vitro tests could be carried out to determine carcinogenesis.

Progress : start = 22/04/80   end = 31/03/81   publication =

Person(s) responsible : Dr Denis Nadeau.

Sponsor(s) : Ministère de l'Energie et des Ressources (MER), Gouvernement du Québec, Québec, QC, Canada G1R 4X7.

Cooperation : Institute for Mineral Industrial Research, Occupational Health and Safety Unit, McGill University; Laboratoire de caractérisation de l'amiante (LCA), Université de Sherbrooke.

Publication(s) on the project :
Nadeau, D. (1981). Cytotoxicité des fibres naturelles et des fibres artificielles. Final Report submitted to the Ministry of Mines and Resources (MER), Quebec.
Nadeau, D., D. Paradis, A. Gaudreau, J.P. Pelé and R. Calvert (1983). Biological evaluation of various natural and man-made mineral fibers: cytotoxicity, hemolytic activity and chemiluminescence study. Environ. Health Perspec. 51:374.

Codes : A.9.a) - A.12 - C.1 - C.2 - C.4 - C.5.a) - C.5.b) - C.7 - D.0 - E.5 - E.6 - E.11 - E.35

Canada

LABORATOIRE DE CARACTERISATION DE L'AMIANTE (LCA),
FACULTE DES SCIENCES, UNIVERSITE DE SHERBROOKE.

Original language : French

Title : **BIOLOGICAL AND PHYSICO-CHEMICAL CHARACTERIZATION OF ASBESTOS FIBRES
AND OTHER FIBROUS MATERIALS.**

Research protocol :

The main task of the LCA is the detailed analysis of fibrous materials.
This is carried out by the Research Division of the Asbestos Institute,
researchers associated with the Research Programme on Asbestos (Programme
de recherche sur l'amiante), various private industries, and/or bodies
concerned with the salubrity of environmental air. The LCA's expertise
lies in the following areas:

a) determination of in vitro biological reactivity:
   - haemolytic potential
   - cytotoxicity on alveolar macrophages
   - genotoxicity

b) physico-chemical characterizations:
   - enumeration and morphometric study (length and diameter)
   - infrared spectroscopy
   - specific surface (BET)
   - thermal analyses (ATD and ATG)
   - zeta potential
   - various chemical analyses

Progress : start = 01/01/83   end = 31/12/85   publication = 01/06/86

Person(s) responsible : Dr Denis Nadeau.

Sponsor(s) : Institut de l'amiante
             Institut de l'amiante, Division de la recherche, Sherbrooke

Cooperation : Institut de l'amiante, Division de la recherche; Programme de
   recherche sur l'amiante de l'Université de Sherbrooke; Laboratoire de
   Biochimie et de toxicologie pulmonaires (URP), Université de Sherbrooke.

Publication(s) on the project :
   Dunnigan, J., D. Nadeau and D. Paradis (1984). Cytotoxic effects of aramid
   fibres on rat pulmonary macrophages: comparison with chrysotile asbestos.
   Toxicol. Lte. 20; 277-282.
   Nadeau, D., L. Fouquette-Couture, J. Khorami, D. Paradis, D. Lane and J.
   Dunnigan (1986). Cytotoxicity of respirable dusts from industrial
   minerals: comparison of two naturally-occuring and two man-made silicates.
   TEES Monograph Series (to be published).
   (cont. p. 193)

Codes : A.0 - A.9.a) - A.3 - A.4 - A.12 - C.0 - C.1 - C.3 - C.4 - C.5.a) -
   C.5.b) - C.6 - D.0 - E.0 - E.2 - E.5 - E.6 - E.11 - E.22 - E.33 - E.35.

LABORATORY OF CATALYSIS, DEPARTMENT OF CHEMISTRY, CONCORDIA UNIVERSITY.

Original language : English

Title : **SYNTHESIS AND CHARACTERIZATION OF COMPOSITE ZEOLITE/ASBESTOS CATALYSTS AND ADSORBENTS.**

Research protocol :

A procedure has been developed which allows growth of a zeolite in chrysotile (asbestos) fibers which have previously had magnesium partially leached out. The synthesis is now extended to all zeolites of industrial interest. The resulting materials will be tested in several catalytic and adsorption processes.

Progress : start = 01/08/83   end =   publication =

Person(s) responsible : Dr. Raymond Le Van Mao

Sponsor(s) : The Canadian Asbestos Institute,
             The Canadian Mines and Energy Technologies (CANMET)

Cooperation :

Publication(s) on the project :

1) R. Le Van Mao and P.H. Bird, US Patent 4,511,667 (1985) and Canadian Patent 1,195,311 (1985)
2) R. Le Van Mao, P. Levesque, B. Sjiariel and PH.H. Bird, Can. J. Chem. 63 (1985)
3) R. Le Van Mao, P. Levesque, B. Sjiariel, Can. J. Chem. Eng., in press

Codes : A.3 - B.14 - C.1 - D.13

ONTARIO MINISTRY OF LABOUR.

Original language : English

Title : **PULMONARY FUNCTION IN ASBESTOS-CEMENT WORKERS: A DOSE-REPONSE STUDY**

Research protocol :

A new measure of asbestos dose, namely residence-time weighted exposure, which was successfully used by the author to model the risk of radiographic abnormalities in asbestos cement workers (BJIM 1985;42;319) has been applied to the study of pulmonary function in the same cohort. It was found that once again this measure provided estimates of risk that were independent of latency and permitted modeling of dose-response relationships without the explicit use of time.

Progress : start =   end =   publication = 1986

Person(s) responsible : Dr. Murray Finkelstein.

Sponsor(s) :

Cooperation :

Publication(s) on the project :
1) Am Rev Respir Dis 1984; 129:754.
2) BJIM 1985;42;319.
3) BJIM (in press).

Codes : A.10 - A.11 - C.1 - C.2

Canada

ONTARIO RESEARCH FOUNDATION.

Original language : English

Title : **ASBESTOS FIBRE MEASUREMENTS IN BUILDING ATMOSPHERES.**

Research protocol :

Workshop to evaluate the available data on indoor asbestos fibre concentrations based on a critical review of the analytical methods, and to draw conclusions concerning appropriate methods of sampling and analysis for airborne asbestos in building atmospheres.

Progress : start = 02/1985   end = 12/1985   publication = 12/1985

Person(s) responsible : Dr. E.J. Chatfield, O.R.F.
                        Ms. M.E. Meek, Health & Welfare.

Sponsor(s) : Health and Welfare Canada.

Cooperation : Health and Welfare Canada.

Publication(s) on the project : Proceedings to be published December 85.

Codes : A.12 - B.3 - D.3.a)

PROGRAMME DE RECHERCHE SUR L'AMIANTE, UNIVERSITE DE SHERBROOKE. (1)

Original language : French

Title : **MOLECULAR ASPECTS OF ASBESTOS BIOLOGICAL REACTIVITY**

Research protocol :

The project is aimed at evaluating the nature and density of various types of surfaces sites which may be involved in the biological reactivity of chrysotile asbestos. The approach followed comprises binding studies for various types of molecular probes and selective surface chemical derivatization of the chrysotile. Parallel evaluation of the biological reactivity of the fibres is carried out from hemolytic activities.

Progress : start = 01/11/86   end = 01/11/89   publication =

Person(s) responsible : Carmel Jolicoeur.

Sponsor(s) : Natural Sciences and Engineering Research Council.

Cooperation :

Publication(s) on the project :

Codes : A.3 - A.9 - B.0 - C.1 - C.3

Canada

PROGRAMME DE RECHERCHE SUR L'AMIANTE, UNIVERSITE DE SHERBROOKE. (2)

Original language : French

Title : **ADSORPTION OF CARCINOGENIC MOLECULES ON PARTICULATE MATERIALS.**

Research protocol :

Fibrous and particulate materials of respirable size form a large and extremely diversified category of industrial materials. The biological activity of the particles may be small in some cases or large in others, but it is considerably amplified when the particles adsorb carcinogenic molecules. The problem, therefore, is to determine which particles possess active sites for the adsorption of carcinogenic molecules.

The purpose of the project is, therefore:

- to determine which industrial fibrous and non-fibrous particles have an adsorption potential vis-à-vis carcinogenic molecules.

- to determine the adsorption isotherms by chromatographical and thermo-gravimetrical techniques.

Progress : start = 01/01/84   end = 01/01/87   publication =

Person(s) responsible : Hugues Ménard.

Sponsor(s) : Société nationale de l'amiante / L'Institut de l'amiante /
             Institut de recherche en santé et en sécurité du travail du Québec.

Cooperation :

Publication(s) on the project :

Codes : A.3 - A.11 - C.0 - E.0

PROGRAMME DE RECHERCHE SUR L'AMIANTE, UNIVERSITE DE SHERBROOKE. (3)

Original language : French

Title : **CHEMICAL METHODS FOR QUALITY CONTROL OF CHRYSOTILE ASBESTOS FIBRES.**

Research protocol :

A broad research program has been undertaken to devise automated sequential chemical analysis in order to rapidly identify the physico-chemical properties of asbestos fibers in aqueous slurries. The work is carried out in order to provide rapid means of identifying characteristic differences between asbestos fibers of different geological origin, or following various types of physical or chemical treatment.

Progress : start = 01/01/86   end = 01/01/89   publication =

Person(s) responsible : Carmel Jolicoeur.

Sponsor(s) : Natural Sciences and Engineering Research Council.

Cooperation : Société nationale de l'amiante, L'Institut de l'amiante.

Publication(s) on the project :

Codes : A.3 - A.6.b) - B.0 - B.9 - C.1 - C.3

39

PROGRAMME DE RECHERCHE SUR L'AMIANTE, UNIVERSTE DE SHERBROOKE. (4)

Original language : English

Title : **COMBINED THERMOGRAVIMETRY AND FOURIER TRANSFORM INFRARED
SPECTROSCOPY (FTIR) TECHNIQUES FOR GAS EVOLUTION ANALYSIS.**

Research protocol :

The usefulness of thermogravimetry has been amply demonstrated for a
wide variety of material analysis applications. In many instances,
however, additional information is required for adequate characterization of the sample and its thermal decomposition behavior. In this
respect, the analysis of evolved gases, or condensed liquids, has proven
a highly useful approach. Among the various physical methods used for
analysis of the thermal degradation products, infrared spectroscopy has
often been found very powerful, being versatile, rapid and widely
accessible.

In this study, we report a simple new approach in which the evolved
gases and condensed liquids from the thermal decomposition of various
products are recuperated, respectively in an infrared gas cell and on a
PVC membrane filter. The gaseous components were analysed by
transmission FTIR, and the condensed liquid products were examined
directly on the PVC membrane by FTIR in the internal reflexion mode. The
technique was used, for example, to examine the pyrolysis products
(gases and liquid) of Koberit, a proposed substitute for asbestos. The
method was also applied to the study of chemically derivatized asbestos
materials in attempt to unravel the surface chemical modifications.

Progress : start = 06/84   end = 08/85   publication = 1986. In press

Person(s) responsible : J. Khomari, G. Chauvette, A. Lemieux, J. Ménard and
C. Jolicoeur.

Sponsor(s) : The Asbestos Institute.

Cooperation :

Publication(s) on the project :

Codes : A.16 - B.9 - C.0 - E.0

40

PROGRAMME DE RECHERCHE SUR L'AMIANTE, UNIVERSITE DE SHERBROOKE.(5)

Original language : French

Title : **QUANTITATIVE MEASUREMENT OF RESPIRABLE CHRYSOTILE ASBESTOS DUST BY INFRARED SPECTROSCOPY WITH INTERNAL REFLECTION.**

Research protocol :

A technique has been developed for the quantitative measurement of respirable chrysotile asbestos dust by using an adsorption band of 303 cm^{-1} on infrared spectra, recorded by the internal reflection method. This technique is quick, simple and enables us to detect quantities of chrysotile asbestos at the level of a microgramme per centimetre squared ($\mu g/cm$).
In this spectral region (303 cm^{-1}), the possibility of bands connected with the presence of interference is very low. The asbestos dusts are collected on a PVC membrane at varying intervals of time and in accordance with the well defined experimental conditions of a controlled dust chamber (1,2). The dust chamber arbitrarily reproduces the atmospheric conditions of a workshop contaminated with asbestos. Standard curves are established between the intensity (absorbance) of the band at 303 cm^{-1}, determined by internal reflection IR spectroscopy, and the concentration of respirable chrysotile asbestos fibres, determined by atomic adsorption spectrometry. The results obtained are confirmed by the method currently used to evaluate the concentration of respirable fibres, i.e. optical microscopy, and by transmission electron microscopy. The results obtained by internal reflection spectroscopy demonstrate the reliability of this technique and, in particular, its application with low quantities of asbestos, which the current method (optical microscopy) is sometimes incapable of evaluating, or if transmission electron microscopy is used, the procedure is very long and expensive to perform.

Progress : start = 1983   end = 1985   publication = 1986

Person(s) responsible : C. Létourneau, J. Khorami, F. Kimmerle et C. Jolicoeur.

Sponsor(s) : Institut de l'amiante.

Cooperation :

Publication(s) on the project :

Codes : A.12 - C.1 - B.5

Canada

PROGRAMME DE RECHERCHE SUR L'AMIANTE, UNIVERSITE DE SHERBROOKE. (6)

Original language : English

Title : **PHYSICO-CHEMICAL CHARACTERIZATION OF ASBESTOS AND ATTAPULGITE MINERAL FIBERS BEFORE AND AFTER TREATMENT WITH PHOSPHOROUS OXYCHLORIDE.**

Research protocol :

Among the hydrous silicates belonging to the serpentine, amphibole, and clay mineral families, chrysotile, crocidolite and attapulgite fibers reacted the strongest with phosphorous oxycloride. While the reactivity of crocidolite was linked mainly to the sodium cations of its structure, the reactivities of chrysotile and attapulgite correlated best with their high content in hydroxyl groups. The thermogravimetric curves of chrysotile and attapulgite revealed significant modifications of their dehydroxylation profiles. Infrared spectra and specific surface measurements confirmed that, most probably, the phosphorylation process created: (a) a phosphorous coating at the surface of the chrysotile fibers, and (b) an obstruction of the pores by phosphorous compounds with the attapulgite fibers.

Progress : start = 1983   end = 1985   publication = 1986

Person(s) responsible : J. Khorami and D. Nadeau.

Sponsor(s) : The Asbestos Institute.

Cooperation :

Publication(s) on the project :

Codes : A.5 - B.9 - C.3

Canada

SCHOOL OF OCCUPATIONAL HEALTH, McGILL UNIVERSITY.

Original language : French

Title : **BIOLOGICAL MONITORING AND CARCINOGENICITY OF ASBESTOS IN THE WORKPLACE.**

Research protocol :

Studies on tissue burden data in asbestos textile workers and correlation with lung cancer incidence.

Progress : start = 01/07/83   end = 31/03/86   publication =

Person(s) responsible : Dr. J.C. McDonald/Dr. P. Sébastien.

Sponsor(s) :   Institut de l'amiante.

Cooperation :

Publication(s) on the project :

Codes : A.10 - A.11 - C.1

PNEUMOCONIOSIS RESEARCH UNIT,
SCHOOL OF PUBLIC HEALTH, UNIVERSITY OF MEDICAL SCIENCES. (1)

Original language : Chinese

Title : **AN EXPERIMENTAL STUDY ON THE CARCINOGENIC EFFECTS OF ASBESTOS IN CHINA.**

Research protocol :

This paper outlines the results of an experimental study on the carcinogenic activity of asbestos in China.
Since 1982, animal experiments have been carried out. Rats and monkeys
were intrapleurally inoculated with chrysotile and crocidolite produced in
different mines. At the same time, UICC asbestos was also used.
This study includes the following subjects:
1. The technique of preparing asbestos fibers for experimental use.
2. The technique of intrapleural inoculation for inducing pleural
   mesothelioma.
3. Analysing the composition of asbestos, fiber dimension and X-ray
   Spectrometry.
4. The construction of the animal model of pleural mesothelioma in rats
   and monkeys.
5. Comparing the incidence of induced mesothelioma using Chinese asbestos
   (5 types chrysotile and 3 types crocidolite) with that of UICC
   asbestos.
6. Contrasting the manifestations (micro focus 0.3 x 0.3 mm) of animal
   mesothelioma as observed from X-rays with that observed in pathologic
   changes.
7. Analysing the characteristics of animal pleural mesothelioma by
   Electron Microscope.
8. Types of pathological changes of pleural mesothelioma in rats.

Progress : start = 1983   end = 1985   publication = 1986

Person(s) responsible : Liu Xueze and Luo Suqiong.

Sponsor(s) : Science Fund of the Chinese Academy of Sciences.

Cooperation : Institute of Cancer Research of West China University of Medical
              Sciences.

Publication(s) on the project :
    1. Journal of West China University of Medical Sciences.
    2. Chinese Journal of Industrial Hygiene and Occupational Diseases.

Codes : A.9(b) - B.7 - C.1 - C.2

PNEUMOCONIOSIS RESEARCH UNIT,
SCHOOL OF PUBLIC HEALTH, UNIVERSITY OF MEDICAL SCIENCES. (2)

Original language : Chinese

Title : **RESEARCH INTO THE CARCINOGENIC DANGERS OF CROCIDOLITE IN CHINA.**

Research protocol :

This paper presents the results of research into the carcinogenic dangers of crocidolite in China.
In 1983, contamination of environmental crocidolite was found in certain regions of China. And some cases of malignant mesothelioma associated with asbestos exposure were found.
This study includes the following subjects:
1. An environmental study on crocidolite contamination.
2. The epidemiologic study on carcinogenic dangers with a cross-section study and retrospective cohort study.
3. The characteristics of radiological manifestations of pleural plaques and calcifications.
4. The clinico-pathologic and X-ray study of some malignant mesotheliomas.
5. The occurrence of endemic asbestosis among farmers.
6. The video taping of environmental contamination of crocidolite and autopsies of a few mesothelioma cases.

Progress : start = 1984   end = 1985   publication = 1986

Person(s) responsible : Liu Xueze and Wang Zhiming.

Sponsor(s) : Science fund of the Educational Committee of the PRC.

Cooperation : Department of Pathology of West China University of Medical Sciences.

Publication(s) on the project :

1. Journal of West China University of Medical Sciences.
2. Chinese Journal of Industrial Hygiene and Occupational Diseases.

Codes : A.9 - A.10 - A.12 - B.5 - B.6 - B.7 - C.2

INSTITUTE OF TUBERCULOSIS AND RESPIRATORY DISEASES.

Original language : Czech

Title : **INFLUENCE OF CHRYSOTILE ON THE DEVELOPMENT OF LUNG MALIGNANCIES IN ASBESTOS MILLING AND MINING WORKERS IN SLOVAKIA.**

Research protocol :

A prospective clinical and epidemiological study, which examines the occurrence of lung cancer and malignant mesothelioma in workers in the mining and milling of asbestos/chrysotile. The work was started in 1981 and should be finished in 1990. 150 employees are being regularly examined: personal and professional history, the length of exposure to asbestos, the asbestos load measured by semi-quantitative investigation of the sputum, the length of time and intensity of smoking, chest X-ray in PA and lateral projection, lung function. The relations between the above-mentioned data are monitored. Half-way through the study - 1985 - only one case of lung carcinoma had been noticed (with asbestosis) after 30 years of continuous exposure. The average length of exposure of the whole cohort is 18 years. No malignant mesotheliomas or malignances of other sites have been noticed. Partial results suggest probably a low carcinogenic effect of chrysotile mined in a small area of Slovakia. Final conclusions will be stated after 10 years' research, in 1990.

Progress : start = 01/06/81   end = 30/06/90   publication =

Person(s) responsible : MUDr. Jurikovič Milan CSc. 05984 Vyšné Hágy, CSSR.

Sponsor(s) :

Cooperation : MUDr. Rovenský Eugen, 05801 Kvetnica, CSSR.

Publication(s) on the project :

Codes : A.2 - A.11 - B.1 - B.6 - C.1

46

INSTITUTE OF OCCUPATIONAL HEALTH. (1)

Original language : English

Title : **ASBESTOSIS, THE NERVOUS SYSTEM AND CANCER.**

Research protocol :

The Institute of Occupational Health began a follow-up study of 115 asbestosis patients in 1979. Neurological and neuro-physiological examinations were the focal point of the study. Disorders of the nervous system are often early signs of cancer developing in the body. Such disorders can be detected before the cancer has become manifest. Asbestosis patients were taken under continuous scrutiny of the nervous system because they are known to have a high risk of cancer, lung cancer in particular. The referents comprise patients with fibrosing alveolitis and silicosis, a disease similar to asbestosis but not asbestos-induced, and also a group of patients with gynecological cancer.

The preliminary results show that asbestosis patients have an exceptionally high prevalence of nervous system dysfunction; 17 percent of them had peripheral neuropathy; 17 percent had involvement of the central nervous system; and 22 percent had both peripheral neuropathy and disturbances in the central nervous system. These prevalences were clearly higher than those for the referents, only 10 percent of whom had peripheral neuropathy. Patients with rapidly progressing asbestosis had fewer nervous system disorders in clinical examinations than did other asbestosis patients, but they reported more subjective symptoms.

(cont. p. 193)

Progress : start = 01/01/79   end = 31/12/86   publication = 31/12/87

Person(s) responsible : J. Juntunen and M.S. Huuskonen.

Sponsor(s) :

Cooperation : University Hospital of Helsinki; Laboratory of Clinical Immunology, Helsinki.

Publication(s) on the project :
Juntunen J., Huuskonen M.S., Matikainen E., Kemppainen R., Suoranta H., Tukiainen P., Korhonen O., Järvisalo J. & Partanen T.: Asbestosis, the nervous system and cancer. Ann. Acad. Med. 13 (1984): 2,353-360.
Jarvisalo J. Juntunen J, Huuskonen M.S., Kivistö H. & Aitio A.: Tumor markers and neurological signs in asbestosis patients. Am.j.industr.med. 6 (1984) 160-166.
(cont. p.193)

Codes : A.9.b) - A.11 - B.6

INSTITUTE OF OCCUPATIONAL HEALTH. (2)

Original language : English

Title : **HEALTH EFFECTS OF EXPOSURE TO WOLLASTONITE.**

Research protocol :

> Wollastonite is a naturally occurring acicular or fibrous calcium silicate used in ceramics and sometimes as a substitute for asbestos. Wollastonite fibers are rather similar in form, length, and diameter to the fibers of amphibole asbestos. Dust measurements in a Finnish quarry and flotation plant showed high concentrations of dust and fiber in some operational stages.

> A total of 46 men who had been exposed in limestone-wollastonite quarry for at least ten years, over 20 years on average, were clinically examined. Radiographs revealed slight lung fibrosis in 14 men and slight bilateral pleural thickening in 13 men. Their sputum specimens were normal. Spirometry and nitrogen single breath test indicated the possibility of small airways diseases.

> A slight elevation in the serum activities of ACE, LZM, and lysosomal enzymes (NAG and GLU) was found when the quarry workers with radiographic pulmonary changes were compared with workers without those changes. The quarry workers also had a higher level of rheumatoid factors than the Finnish blood donors who were used as referents. This finding is similar to that among asbestos workers.

> A mortality study of 238 quarry workers was nonpositive. However, 30 years after initial exposure, one woman who had been exposed to wollastonite for 20 years had a retroperitoneal malignant tumor which was diagnosed to be a mesenchyoma. This kind of tumor is very rare, and it is difficult to distinguish from mesothelioma. Since only one case was detected, no definite conclusion can be drawn. The follow-up will be continued.

Progress : start = 01/04/81   end =   publication =

Person(s) responsible : M.S. Huuskonen and A. Tossavainen.

Sponsor(s) :

Cooperation : Laboratory of Clinical Immunology, Helsinki.

Publication(s) on the project :
> Huuskonen M.S., Järvisalo J, Koskinen H, Nickels J, Räsänen J & Asp S.: Preliminary results from a cohort of workers exposed to wollastonite in a Finnish limestone quarry. Scand.j.work environ.hlth 9 (1983):2 (Special issue), 169-175.

> (cont. p.193)

Codes : A.9.a) - A.9.b) - A.11 - B.6 - B.7 - C.1 - C.2 - E.35

Finland

INSTITUTE OF OCCUPATIONAL HEALTH. (3)

Original language : English

Title : **CLINICAL FOLLOW-UP OF PATIENTS WITH PNEUMOCONIOSES.**

Research protocol :

Patients with asbestosis and silicosis undergo regular medical examinations at the Institute of Occupational Health. These patients, 155 asbestosis and 144 silicosis patients, were the subjects of a clinical cross-sectional study performed a few years ago and are now being followed up. The pulmonary fibrosis in patients exposed to asbestos may have different pathogenetic mechanisms than the pulmonary fibrosis in patients exposed to silica dust. The aim of the follow-up study is to evaluate these differences.

Clinical and laboratory variables will be evaluated in order to assess the activity of the fibrotic process. The laboratory studies include the determination of some enzyme activities in serum (angiotensin-converting enzyme, elastase, adenosine deaminase, fibronectin, some lysosomal enzymes) and the determination of the concentrations of some enzymes which participate in collagen biosynthesis. The clinical examinations include the evaluation of respiratory symptoms, radiography, and pulmonary function tests. Nervous system involvement is being investigated as a separate study.

The presence of tumor-associated antigens and markers (such as CEA, beta2-microglobulin, ferritin, and speudouridine) and the concentration of acute phase reactant in serum are also being determined. The immunological parameters to be determined are Latex, Waaler-Rose, antinuclear antibodies, and complements of C3 and C4. Cytological studies of sputum will also be performed.

The mortality of patients with asbestosis is being followed up. The results of the study will be reported by the end of 1987.

Progress : start = 01/01/79   end = 31/12/86   publication = 31/12/87

Person(s) responsible : H. Koskinen.

Sponsor(s) :

Cooperation : Meltola Hospital, Karjaa; World Health Organization, Copenhagen, Denmark; Laakso Hospital, Helsinki; University of Oulu; Laboratory of Clinical Immunology, Helsinki; University Hospital of Helsinki.

Publication(s) on the project :

Codes : A.9.b) - A.11 - B.6

Finland

INSTITUTE OF OCCUPATIONAL HEALTH. (4)

Original language : English

Title : **LYSOSOMAL ENZYMES AND ANGIOTENSIN CONVERTASE IN OCCUPATIONAL LUNG DISORDERS.**

Research protocol :

The serum activities of both angiotensin converting enzyme (ACE, EC 3.4.15.1) and two lysosomal enzymes, β-N-acetylglucosaminidase (NAG, EC 3.2.1.30) and β-glucoronidase (GLU, EC 3.2.1.31) were analyzed in 70 asbestosis patients. A total of 127 reference workers from the food industry without exposure to asbestos showed lower levels of these enzymes than did asbestosis patients, and among these age and sex did not correlate with the levels of the enzymes. On the group basis, the levels of any of these enzymes in serum of patients with asbestosis revealed no correlation with the radiographic profusion score of small opacities and neither did their levels differ in patients having radiographically progressive or nonprogressive asbestosis, except for the decreased NAG level among the progressive subjects. No explanation was found for this decrease.

Progress : start = 1980   end =   publication = 1986

Person(s) responsible : M.S. Huuskonen, J. Järvisalo, H. Koskinen & H. Kivistö.

Sponsor(s) :

Cooperation :

Publication(s) on the project :

Codes : A.9.b) - A.11 - B.6

France

CENTRE DE RECHERCHES SUR LA PHYSICO-CHIMIE DES SURFACES SOLIDES (CNRS).(1)

Original language : French

Title : **GRINDING OF ASBESTOS**.

Research protocol :

A previous study showed that the conditions (liquid) in which grinding takes place, considerably affect the morphology of chrysotile fibres (Canada) ground in a ball-grinder. In particular, defibrillation is observed in aquatic conditions and segmentation in organic conditions. These results were explained on the basis of existing theories. The observations have subsequently been checked with other chrysotiles (S. Africa) and the tendencies observed with the chrysotiles have also been found in the behaviour of crocidolite and asbestos.

Progress : start = 1980   end =1986   publication = 1986

Person(s) responsible : P. Roland.

Sponsor(s) : CNRS.

Cooperation :

Publication(s) on the project :

E. Papirer, P. Roland: Grinding of chrysotile in hydrocarbons, alcohol and water. Ch. Ch. Miner., 29 (3) 161 (1981).
E, Papirer, JB Donnet and P. Roland: Comminution of asbestos: role of the liquids used as grinding aids, Tribology Series 7, J.M. Georges Ed., Elsevier p. 449-458 (1982).
E. Papirer: The effect of filler shape on the mechanical properties of a reinforced vulcanizate: The SBR-ground asbestos system. J. Polym Sci (Polym Chem Ed) 21 2833 (1983)

Codes : A.0 - B.3 - C.1 - C.2 - D.6

France

CENTRE DE RECHERCHES SUR LA PHYSICO-CHIMIE DES SURFACES SOLIDES (CNRS).(2)

Original language : French

Title : **STABILITY OF CHRYSOTILE IN ALKALINE CONDITIONS,
AT A TEMPERATURE BETWEEN 120 and 180°C.**

Research protocol :

A study of the chemical corrosion in autoclaves of chrysotile asbestos
fibres in strong alkaline conditions (between 20 and 40% by weight of KOH)
in the temperature range: 120-180°C, then an attempt to find a
stabilisation process for asbestos chrysotile in the same conditions.

Progress : start = 13/09/83   end = 13/09/86   publication =

Person(s) responsible : Mr. Bernard Siffert.

Sponsor(s) : Electricité de France (E.D.F.)

Cooperation : Laboratoires E.D.F. - Saint-Denis, France

Publication(s) on the project :

Codes : A.3 - B.4 - C.1 - D.2.b) - D.9

France

CLINIQUE TOXICOLOGIQUE, HOPITAL FERNAND-WIDAL.

Original language : French

Title : **THE EFFECT OF EXPOSURE TO ASBESTOS DUSTS ON THE RESPIRATORY FUNCTION**

Research protocol :

Survey of 717 workers exposed to asbestos dusts and comparison with a control group. One of the aims was the early detection of respiratory function disorders before the appearance of clinical and radiological signs.

Presentation of the subjects; parameters measured; statistical study; comments on the results and how they compare with information in existing literature.

It would seem that a significant number of exposed workers suffer from some form of respiratory function disorder, even when no signs appear on X-rays. The disorders only come to light when flow volume curves, and in particular FEF 75%, are studied. This method seems more sensitive than conventional spirographic methods.

Progress : start = 1979   end = 1984   publication =

Person(s) responsible : M. Danan, V. Massari, O. Diamant-Berger, R. Badin, S. Dally, P.E. Fournier, A.-J. Valleron.

Sponsor(s) :

Cooperation : Unité de Recherches Biomathématiques et Biostatistiques, INSERM U 263, Université Paris 7.

Publication(s) on the project :

"The effect of exposure to asbestos dusts on the respiratory function" Cahier de notes documentaires de l'INRS, 2è trim. 1985, n° 119.

Codes : A.9 - A.10

France

ECOLE SUPERIEURE DE L'ENERGIE ET DES MATERIAUX, UNIVERSITE D'ORLEANS.

Original language : French

Title : **COMPARATIVE STUDY OF DNA ADSORPTION ON CHRYSOTILE AND CHRYSOPHOSPHATE**

Research protocol :

Deoxyribonucleic acid (DNA) adsorption on one sample of Canadian chrysotile and two samples of chrysophosphate has been performed using chemical analysis and surface method of analysis (spectrometry of photo-electrons, XPS). Chrysophosphate was obtained using a phosphatation process applied to chrysotile fibres; one chrysophosphate sample has been opened in a Pallmann device.

XPS analysis of the samples before DNA adsorption has shown that some surface phosphorus was solubilised following suspension of the fibres in distilled water. The amount of DNA adsorbed at the surface of the chrysophosphate sample was lower than that adsorbed at the chrysotile surface (12.6 ± 7 µg/mg and 71 ± 5.4 µg/mg respectively). In addition, the opened sample adsorbed on intermediary DNA amount (27.7 ± 5.6 µg/mg) suggesting that the phosphatation process has not reached the all bulk fibres.

The results indicate that DNA adsorption is a sensitive method to appreciate the fibre surface state. Elsewhere, it seems that the phosphatation process masks or blocks the superficial sites reactive in the DNA adsorption.

Progress : start = 01/01/84   end = 01/02/85   publication = 1986
                         submitted to "Canadian Journal of Chemistry"

Person(s) responsible : J.C. Touray, professeur.

Sponsor(s) : SNA, Quebec Canada; CNRS & INSERM.

Cooperation : CHU Henri Mondor, INSERM U139.

Publication(s) on the project :

Codes : A.0 - C.1 - D.1

I.N.S.E.R.M. UNITE 139. (1)

Original language : French

Title : **RELATIONSHIP BETWEEN PHYSICO-CHEMICAL PROPERTIES OF MINERAL FIBRES AND THEIR BIOLOGICAL EFFECTS.**

Research protocol :

The production of free radicals by mineral fibres (asbestos and attapulgite) using electronic paramagnetic resonance (spin trapping).

Sorptive properties of the fibres (attapulgite and modified chrysotile) vis-à-vis biological macromolecules: DNA, phospholipids and proteins. Analysis of adsorption, by chemical methods and photoelectron spectrometry (surface analysis).

Progress : start = 1985   end = 1986   publication =

Person(s) responsible : M.C. Jaurand, M. Perderiset.

Sponsor(s) : France: INSERM, Ministère de la Recherche et de l'Industrie, Rhône-Poulenc; Canada: Société Nationale de l'Amiante.

Cooperation : H. Pezerat (CNRS, Université Paris VI)
              J.C. Touray (Université d'Orléans, CNRS).

Publication(s) on the project :

Codes : A.9.a) – C.1 – C.3 – C.5.a) – E.6

55

France

I.N.S.E.R.M. UNITE 139. (2)

Original language : French

Title : **PHENOTYPE, KARYOTYPE AND TUMORIGENICITY OF A RAT MESOTHELIAL CELL
LINE OBTAINED FROM A CHRYSOTILE-INDUCED MESOTHELIOMA.**

Research protocol :

Characteristics of the line in culture: analysis of the cell cycle in flow
cytometry, growth in semi-solid conditions, production of growth and
transformation factors.

Analysis of the karyotype of the line (R banding).

Tomorigenicity determined by injection of cells into the athymic mouse.
Ultrastructural analysis of tumors induced and comparison with the ultra-
structure of the original cells.

Progress : start = 1985   end =1987   publication =

Person(s) responsible : M.C. Jaurand, F. Fleury, F. Levy, J. Bignon.

Sponsor(s) : INSERM.

Cooperation : B. Dutrillaux (CNRS), D. Barritault (CNRS).

Publication(s) on the project :

Codes : A.9.a) — C.1

France

I.N.S.E.R.M. UNITE 139. (3)

Original language : French

Title : **IN VITRO CLASTOGENIC AND TRANSFORMING POTENCY OF MINERAL FIBRES ON RAT PLEURAL MESOTHELIAL CELLS**

Research protocol :

Study into DNA repair (synthesis not programmed) and the inducing of chromosome anomalies (chromosome aberrations and exchanges of sister chromatids) by various types of mineral fibres (attapulgite, modified chrysotile, asbestos) and research into the mechanisms involved in the formation of these anomalies.

Study of phenotypic transformation (loss of contact inhibition, growth in agar) in a long-term transformation model. The cultures are repeatedly treated for several weeks with the fibres under test (asbestos and attapulgite). A comparison is drawn with the results obtained from the short-term transformation models.

Progress : start = 1985   end =   publication =

Person(s) responsible : M.C. Jaurand.

Sponsor(s) : France: INSERM, Ministère de la Recherche et de l'Industrie; Canada: Société Nationale de l'Amiante.

Cooperation : I. Emerit (CNRS).

Publication(s) on the project :

- M.J. Paterour, J. Bignon, M.C. Jaurand, Carcinogenesis, 1985, 6, 523-529
- M.C. Jaurand, L. Kheuang, L. Magne, J. Bignon, Mutation Research (in press)
- M.C. Jaurand, A. Renier, A. Gaudichet, L. Kheuang, L. Magne, J. Bignon, NY Acad Sci. (in press).

Codes : A.9.a) - C.1 - C.3 - C.5.a) - E.6

France

I.N.S.E.R.M. UNITE 139. (4)

Original language : French

Title : **CARCINOGENIC AND FIBROGENIC POTENCY OF MODIFIED ASBESTOS FIBRES**

Research protocol :

Anatomo-pathological studies of consecutive lesions in animals exposed to chrysotile asbestos fibres modified by chemical treatment (phosphatation).

Carcinogenic potency: intrapleural injection of 20 mg of fibre in the rat. Groups of 30 - 36 animals; they are kept for the whole duration of their lives.

Fibrogenic potency: intratracheal instillation in hamsters (2 x 3 mg). Groups of 30 animals. They are killed after 1 month, 6 months or 1 year periods.

Progress : start = 1984   end = 1987   publication =

Person(s) responsible : M.C. Jaurand, J. Fleury, M. Nebut, J. Bignon.

Sponsor(s) : Société Nationale de l'Amiante (Canada) and INSERM (France).

Cooperation :

Publication(s) on the project :

Codes : A.9.a) - C.3

France

I.N.S.E.R.M., UNITE 139. (5)

Original language : French

Title : See research protocol.

Research protocol :

   Biometrological research into asbestos fibres and ferruginous bodies as
   indicators of past exposure to asbestos: lung tissue and bronchoalveolar lavage.

Progress : start =   end =   publication =

Person(s) responsible : Pr. Bignon.

Sponsor(s) :

Cooperation :

Publication(s) on the project :

Codes : A.7 - A.9.a)

France

I.N.S.E.R.M., UNITE 139. (6)

Original language : French

Title : See research protocol.

Research protocol :

      Retrospective study of control cases to analyse risk factors (including exposure to asbestos) of lung cancers in man, by means of a job - exposure matrix.

Progress : start =   end =   publication =

Person(s) responsible : Pr. Brochard.

Sponsor(s) : INSERM; Université Paris XII.

Cooperation :

Publication(s) on the project :

Codes : A.9.a) - A.10

France

I.N.S.E.R.M., UNITE 139. (7)

Original language : French

Title : See research protocol.

Research protocol :

    Epidemiological study on the basis of the French mesothelioma register,
comparing subjects exposed to asbestos with those not exposed.

Progress : start =   end =   publication =

Person(s) responsible : Pr. Bignon, Pr. Brochard.

Sponsor(s) : Ministry of Public Health, France.

Cooperation :

Publication(s) on the project :

Codes : A.10

France

I.N.S.E.R.M., UNITE 139. (8)

Original language : French

Title : See research protocol.

Research protocol :

    Epidemiological study on the basis of the French mesothelioma register,
    comparing subjects with calcified pleural plaques who have been exposed
    to asbestos with those not exposed.

Progress : start =   end =   publication =

Person(s) responsible : Pr. Bignon, Pr. Brochard.

Sponsor(s) : INSERM.

Cooperation :

Publication(s) on the project :

Codes : A.10

France

I.N.S.E.R.M., UNITE 139. (9)

Original language : French

Title : See research protocol.

Research protocol :

      Longitudinal epidemiological study of a cohort of permanent workers at the University of Paris VII, who have lived or still live in premises sprayed with asbestos.

Progress : start =   end =   publication =

Person(s) responsible : Pr. Bignon, Pr. Brochard.

Sponsor(s) :

Cooperation : Unité 170 de l'INSERM (Denis Hemon, Sylvaine Cordier) and occupational medicine team of Prof. Proteau.

Publication(s) on the project :

Codes : A.10

France

INSTITUT NATIONAL DE RECHERCHE SUR LA SECURITE (INRS). (1)

Original language : French

Title : **OCCUPATIONAL EXPOSURE TO FIBROUS MATERIALS: STUDY OF THE DIMENSIONAL CHARACTERISTICS AND MEASURE OF POLLUTION LEVELS**

Research protocol :

 The problems connected with the use of asbestos are leading to the increasing use of substitute fibre materials. Little is known about the toxicity of these materials. One of the parameters enabling this to be investigated is the measurement of the dimensional characteristics of the fibres. The metrology and identification of the fibres are being studied (in particular those of composite materials). The metrology of the fibres in certain industries which use this type of material will then be carried out (measuring of concentrations and granulometry).

Progress : start = 01/09/85 end = 31/12/88 publication =

Person(s) responsible : M.E. Kauffer.

Sponsor(s) : Caisse Nationale de l'Assurance Maladie.

Cooperation :

Publication(s) on the project :

Codes : A.7 - B.3 - B.5 - C.4 - C.5.a) - C.5.b)

France

INSTITUT NATIONAL DE RECHERCHE SUR LA SECURITE (INRS).(2)

Original language : French

Title : **STUDY OF THE PULMONARY IMPAIRMENT OF WORKERS ENGAGED IN MAN-MADE MINERAL FIBER PRODUCTION.**

Research protocol :

This study is to be an original contribution to the international study launched by the WHO in 1975.
It consists of a clinical examination and detailed study of the respiratory function of workers in 4 French companies which produce artificial mineral fibres. A full medical questionnaire has been sent to them and will be used as a basis.
A retrospective epidemiological study, covering 10 years, on cancer incidence, is currently being carried out.
Metrological studies and analyses of the atmosphere in places of work have been carried out.

Progress : start = 01/01/83   end = 31/12/86   publication =

Person(s) responsible : Dr. J.J. Moulin.

Sponsor(s) : Caisse Nationale de l'Assurance Maladie.

Cooperation : INSERM U14 - Société St Gobain - IUMT Hte Normandie.

Publication(s) on the project :

Codes : A.5 - A.10 - A.11 - B.3 - B.5 - C.4

France

LA 10 - CNRS & IOPG.

Original language : French

Title : **SEARCH AND EVALUATION OF RAW MATERIALS USED IN ROCKWOOL MAKING.**

Research protocol :

Regional survey of rocks possibly suitable for the production of this material.

Optimisation of treatments leading to the desired finished product, on the basis of geochemical and experimental studies.

Progress : start =   end =   publication =

Person(s) responsible : Jacques Kornprobst, Pierre Boivin, Patrice Foury.

Sponsor(s) : Rockwool Isolation (SA).

Cooperation :

Publication(s) on the project : Confidential.

Codes : A.1 - A.3 - A.4 - A.5 - A.13 - B.1 - B.4 - B.12 - C.7 - D.3.a) - D.3.b) - D.3.c) - E.11

France

LABORATOIRE D'ETUDE DES PARTICULES INHALEES (LEPI). (1)

Original language : French

Title : **ANALYTICAL MICROSCOPY FOR THE ASSESSMENT OF TOTAL FIBRE BURDEN IN HUMAN LUNGS FROM MESOTHELIOMA CASES MATCHED WITH FOUR PATHOLOGICAL SERIES.**

Research protocol :

A case-control study was carried out on twenty cases of pleural mesothelioma matched by age and sex with four equivalent pathological series comprising bronchogenic carcinoma, adenocarcinoma of the lung, lung metastasis and cardiovascular diseases, in order to identify and quantify the total asbestos fibre burden of these patients. Asbestos bodies were counted using light microscopy on standardized samples of 25 mg of dry lung tissue. Uncoated fibres, asbestos and non asbestos, were identified and quantified using an analytical transmission electron microscope on standardized samples of 1 µg of dry lung tissue. The validity of these standardized sampling methods was tested statistically. Results showed that a great variety of fibres were found in lung tissue. The number of non asbestos fibres was very similar to that of asbestos fibres. The higher concentrations of asbestos bodies were found in the mesothelioma series. As regards uncoated fibres, the retention of asbestos fibres was also significantly higher in mesothelioma cases. The concentration of chrysotile fibres was quite identical in the five pathological series. Moreover, the retention of amphibole fibres was higher in mesothelioma cases than in controls. The retention of fibres corresponding to the shape criteria defined by Stanton was higher in mesothelioma cases than in controls, particularly for amphibole fibres, especially crocidolite. The predominant retention of amphibole fibres in lung tissue comparaed with that of chrysotile is discussed.

Progress : start = 06/81   end = 09/85   publication = In press

Person(s) responsible : A. Gaudichet.

Sponsor(s) : INSERM Grant n° 810.158.

Cooperation : Dr. A.Y. Delajartre, Laboratoire d'Anatomie Pathologique, Faculté de Médecine de Nantes, 44035 Nantes, France.

Publication(s) on the project :

Codes : A.7 - B.0

LABORATOIRE D'ETUDE DES PARTICULES INHALEES (LEPI). (2)

Original language : French

Title : **ENVIRONMENTAL ASBESTOS AIRBORNE POLLUTION IN CORSICA AND PLEURAL PLAQUES**

Research protocol :

In Corsica, apart from the unexploited Canari chrysotile mine, there are numerous serpentine outcrops with asbestos fibres scattered around the north eastern area of the island, in contrast with the western granitic area. An epidemiological survey based on chest radiograph data showed a relatively higher risk of pleural plaques among inhabitants of the north eastern area compared with those of north western and southern regions. In order to evaluate the hypothesis that this pleural disease was caused by environmental factors, metrologic monitoring was performed, including the measuring of asbestos pollution in air and water. Sampling sites were chosen in four villages in the western area as controls. Results showed that the levels of asbestos airborne pollution observed in the eastern region ranged between 0.1 and 128 ng/m^3, whereas in the western region, the levels never exceeded 2 ng/m^3. Moreover, tremolite fibres were found only in air samples from the eastern villages. No contamination of tap water was demonstrated. In spite of the low level of airborne asbestos pollution in the villages with a high rate of calcified pleural plaques, measurements of air samples are well correlated with the epidemiological data.

Progress : start = 05/82   end = 09/85   publication = in press

Person(s) responsible : A. Gaudichet and C. Boutin.

Sponsor(s) : INSERM Grant n° 810.155.

Cooperation : Pr. C. Boutin, Hôpital Michel Levy, Université d'Aix Marseille, 13006 Marseille, France.

Publication(s) on the project :

Codes : A.7 - A.10 - B.5 - B.6 - C.1 - C.2

LABORATOIRE D'ETUDE DES PARTICULES INHALEES (LEPI). (3)

Original language : French

Title : **INDOOR ASBESTOS AIR POLLUTION FROM FLOOR TILES**

Research protocol :

In 1982, LEPI reported the first measurements of high concentrations of indoor asbestos airborne pollution associated with the wearing out of asbestos floor tiles (Sébastien et al., Science, 1982, 216, 1410-1413). Additional data on floor tiles has been obtained from the study of 18 buildings. The conclusions we have drawn are:
- The wearing of asbestos floor tiles can induce significant indoor asbestos airborne pollution (up to 50 ng/m^3) with no other asbestos contamination source.
- The shortness of chrysotile fibers emitted from floor tiles, seems to be a good criterion to characterize this type of emission.
- Bearing in mind the size of chrysotile fibers in such emissions (0.6 micron), analytical transmission electron microscopy is the only suitable method for revealing indoor asbestos airborne pollution from floor tiles.

Progress : start = 1982   end = 03/84   publication =

Person(s) responsible : G. Dufour and A. Gaudichet.

Sponsor(s) :

Cooperation :

Publication(s) on the project :

Codes : A.7 - B.0 - B.5 - C.1 - C.2 - D.12

Germany (FRG)

BERAL BREMSBELAG GmbH.

Original language : German

Title : **REPLACEMENT OF ASBESTOS IN FRICTION MATERIALS.**

Research protocol :

The general replacement of asbestos in all kinds of friction materials by substitute materials which are less harmful to health during manufacturing, subsequent handling and when actually in use.

Progress : start = 1974   end = ongoing   publication =

Person(s) responsible : Research and Development Director.

Sponsor(s) :

Cooperation :

Publication(s) on the project :

Codes : A.11 - D.4 - E.0

BERUFSGENOSSENSCHAFTLICHES INSTITUT FÜR ARBEITSSICHERHEIT - BIA.

Original language : German

Title : **COUNTING OF ASBESTOS FIBRES AND OTHER FIBRES WITH THE AID OF QUANTITATIVE IMAGE ANALYSIS**

Research protocol :

The standard procedure for measuring the concentration of asbestos fibres involves the visual counting of fibres by means of a phase-contrast microscope. This method is time-consuming and the results dependent on subjective factors. A suitable picture analysis system (both hard and soft-ware) should therefore be designed and tested. This system should make it possible to automatically determine the concentration of asbestos fibres both in conjunction with phase-contrast microscopy and in conjunction with scanning microscopy.

Progress : start = 01/11/82   end = 31/10/86   publication =

Person(s) responsible : Dr. A. Schuetz; Dr. G. Riediger.

Sponsor(s) : Federal Minister of Research and Technology (FRG).

Cooperation :

Publication(s) on the project :

Codes : A.7 - C.0

Germany (FRG)

FORSCHUNGS- UND ENTWICKLUNGSABTEILUNG, HOECHST AG, WERK KELHEIM. (1)

Original language : German

Title : **DOLANIT ACRYLIC FIBRES OF HIGH TENACITY FOR TECHNICAL APPLICATION.**

Research protocol :

The acrylic fibre "DOLANIT" is a substitute for asbestos in fibre cement products. The technological data on this fibre, together with its resistance, behaviour and applications are described in the publication mentioned below. Short staple fibres are the main type covered.

Progress : start = 1979   end =   publication =

Person(s) responsible :

Sponsor(s) :

Cooperation : Ametex AG, (Eternit) CH-8867 Niederurnen/Switzerland.

Publication(s) on the project : Off-print from Chemiefasern/Textilindustrie, 33/85 (December 1983), 839-846, E 98.

Codes : D.1 - E.37

FORSCHUNGS- UND ENTWICKLUNGSABTEILUNG, HOECHST AG, WERK KELHEIM. (2)

Original language : German

Title : **HIGH-TENACITY ACRYLIC FIBRES FOR TECHNICAL END USE.**

Research protocol :

This expands the coverage of the previous study (1), including applications such as filtration, clutches, yarn- and woven-end uses etc. Long staple crimped fibres are described in the publication mentioned below.

Progress : start = 1979   end =   publication =

Person(s) responsible :

Sponsor(s) :

Cooperation : Ametex AG (Eternit), CH-8867 Niederurnen/Switzerland.

Publication(s) on the project : Melliand Textilberichte 66 (1985), June.

Codes : D.2 - D.2.a) - D.2.b) - D.4.b) - D.9 - D.13 - E.37

Germany (FRG)

INSTITUT FÜR MINERALOGIE, RUHR-UNIVERSITÄT BOCHUM.

Original language : German

Title : **RELATIONSHIP BETWEEN DEFECT STRUCTURE AND CRYSTAL PROPERTIES IN CHEMICALLY SIMPLE, SYNTHETIC AMPHIBOLES**

Research protocol :

Chemical simple amphiboles, chosen to provide a representative composi-
tional sample of this complex mineral group, are synthesized via the
hydrothermal method in "cold-seal", piston-cylinder and internally-heated
gas apparatus. Pressures and temperatures of synthesis extend to 30 kbar
and 900°C. All products are characterized by X-ray diffraction, light
microscopy and high-resolution transmission electron microscopy. Other
analytical, spectroscopic and calorimetric methods are planned where
applicable. The purpose of the study is to relate the extent of the
departure from structural ideality to the experimental parameters of
synthesis and to the macroscopic properties of the crystals and their
aggregates.

Progress : start = 01/10/81   end = 31/11/90   publication =

Person(s) responsible : Deutsche Forschungsgemeinschaft.

Sponsor(s) : Dr. M. Czank, Kiel University.

Cooperation :

Publication(s) on the project : Maresch & Czank (Amer. Minaral., $\underline{68}$, 744-753,
1983; Per. Mineral. (Roma), $\underline{52}$, 463-542, 1983).

Codes : A.1 - A.3 - A.4 - A.5 - B.0 - C.2 - C.4

MARTIN MERKEL GmbH & Co KG.

Original language : German

Title : **REPLACEMENT OF ASBESTOS IN BRAIDED PACKINGS.**

Research protocol :

There are now many asbestos substitutes on the market and it would appear that asbestos can already be replaced by a wide range of materials. Some of the technical qualities of the substitutes are clearly superior to those of asbestos. However, past experiments have shown that in many critical applications, when pressure, temperature and aggressive agents are present simultaneously, asbestos can still not yet be replaced.

The purpose of the project is to replace asbestos in certain seals and braided gaskets by substitutes.

This should lead to the total elimination, both in the production of proofing materials and in their use, of the health risks involved in the use of asbestos.

Progress : start = 01/09/84   end = 31/08/87   publication =

Person(s) responsible : Hans-J. Hedrich.

Sponsor(s) : Federal Minister of Research and Development.

Cooperation :

Publication(s) on the project : 1984 Annual Report on the Research Programme "Humanisation of Working Conditions and Environment".

Codes : D.11.b) - E.4 - E.12 - E.13

PULMOLOGISCHE ABTEILUNG, KRANKENHAUS WITTENBERG-APOLLENSDORF.

Original language : German

Title : **EARLY KNOWLEDGE OF LUNG DAMAGE FROM ASBESTOS DUST IN THE 19th CENTURY IN AUSTRIA**

Research protocol :

> In a family magazine, published in 1888/89, I found a report about the visit of Wilhelm II, the German Emperor, to Franz Joseph, the Austrian Emperor, in Vienna 1888. Franz Joseph escorted his guest to the newly built "Burgtheater". The opera had been opened that year. The journalist reporting on the remarkable technical features of the opera, for instance the air-conditioning and the safety equipment, made the astonishing comment: "The use of asbestos has been avoided on the stage, because the dust it produces is injurious to the lungs. Instead, impregnated clothing has been used."

> Research has been started with a view to finding the name and any publication of the Austrian doctor, who informed the constructor of the Opera – von Hasenauer – about the risk of asbestos for the lung.

Progress : start =   end =   publication =

Person(s) responsible : Dr. med. L. Bergmann.

Sponsor(s) :

Cooperation :

Publication(s) on the project :

Codes : A.10 – A.11 – A.16

India

ENVIRONMENTAL POLLUTION RESEARCH CENTRE.

Original language :

Title : **RECENT DATA ON ASBESTOSIS (1983-1985)**

Research protocol :

There were 18 smokers, the mean age was 42.1 years and the mean duration of exposure to asbestos was 13.7 years. Of the 63 subjects, 95.2% had exertional dyspnoea while one had paroxysmal attacks. Cough was present in 84.1%, sputum in 39.7%, chest pain in 31.8%, general weakness in 25.4%, anorexia in 15.9% and haemoptysis in 2.3%. There was a history of treatment for tuberculosis in 8%. Clinically there was clubbing in 3.2%, associated blood pressure in 3.2%. Clinical examination of the lungs revealed râles in 30.2% and rhonchi in 14.3%.

Chest radiographs were abnormal in all cases. In 9.5% there was one zone involvement (corresponding to gr. 1/2), in 41.2% there was two zone involvement, in another 28.5% there was 3-4 zone involvement (corresponding to gr. II) and in 17.5% there was 5-6 zone involvement (corresponding to gr. III). Predominently there were linear scars (49.2%) or punctate deposits (58.7%). There were nodular or micronodular scars in 23.8%, reticular scars in 12.7% and pleural fibrosis in 8%. There was evidence of calcified pleural plaque in 6.3%.

Lung function was studied by spirometry, bronchodilator test and by testing blood gases at rest and on exercise. In only 19% was FVC above 3 lit; in 54% it was within the normal Indian range. FEV_1 was abnormal in 25.4%, improvement after bronchodilator (over 10%) was seen in 14.3%, pO_2 at rest was below 90 mm in 34%, it was below 70 mm in 8.1%, pO_2 on exercise declined significantly in 41.9%. Airway resistance was measured in 37 of these cases and was normal only in one case. We have so far repeated tests on 13 subjects again. There was progress observed in most subjects.

Progress : start = 1983   end = 1985   publication =

Person(s) responsible : Dr. S.R. Kamat.

Sponsor(s) :

Cooperation :

Publication(s) on the project : Dr. S.R. Kamat, Personal Experiences in Diagnosis of Asbestosis; S.P. Shah, S.R. Kamat, A.A. Nahashur, Pattern of Asbestos Workers in Bombay (in Indian Journal of Occupational Health, resp. March & April 1983, pp. 285 & 305).

Codes : A.11 - B.6

Israel

HEBREW UNIVERSITY, DEPARTMENT OF MEDICAL ECOLOGY.

Original language : English

Title : **FOLLOW-UP OF EX-ASBESTOS WORKERS**.

Research protocol :

    1. Assess smoking status, pulmonary function and morbidity in ex-asbestos workers.

    2. Set up and assess benefits of intervention programmes.

Progress : start =   end =   publication =

Person(s) responsible : E.D. Richter, MD, J. Lafaer, MD et al.

Sponsor(s) :

Cooperation :

Publication(s) on the project : Smoking, Morbidity and Pulmonary Function in Ex-Asbestos Workers - Am. J. Ind. Med. - in press.

Codes : A.11 - B.6 - C.0 - D.2 - D.4

78

LABORATORY OF BIOCHEMISTRY AND
LABORATORY OF CELLULAR IMMUNOLOGY, CARMEL HOSPITAL

Original language :

Title : **FOLLOW UP OF ASBESTOS CEMENT WORKERS (CLINIC RADIOLOGY, IMMUNOLOGY).**

Research protocol :

Investigation of various immunologic, biochemical and hematologic parameters related to clinical and radiological background of workers of an asbestos cement plant, emphasis being made on suppressor cell activity and mitogen stimulation.

Standard test radiographs were obtained according to the standards of the International Labour Office and classified following the Guidlines of 1980 revised edition.

The workers examined are compared with a control group of matched male transport employees.

Progress : start =   end =   publication =

Person(s) responsible : Dr. N. Gruener, Lab. of Biochemistry.
                        Dr. N. Lahat, Lab. of Cellular Immunology.

Sponsor(s) : Committee for prevention in Occupational Health, Ministry of Labour, Israel.

Cooperation :

Publication(s) on the project :

Codes : A.3 — A.9 — A.11 — B.6 — D.1

Italy

CENTRO STUDI E RICERCHE SUGLI EFFETTI BIOLOGICI DELLE POLVERI
INALATE, ISTITUTO DI MEDICINA DEL LAVORO, UNIVERSITA DI MILANO. (1)

Original language : Italian

Title : **STUDY ON EARLY LUNG REACTIONS TO THE INHALATION OF MINERAL FIBERS AT THE PLACE OF WORK.**

Research protocol :

As part of a larger investigation on the early pulmonary effects of occupational inhalation of inorganic and organic dusts, we are studying, by bronchoalveolar lavage (BAL), subjects with a history of past or present occupational exposure to asbestos or man-made mineral fibers, with or without signs of parenchymal or pleural disease. On BAL fluid we are performing (1) cytological studies (total and differential cell counts, study of cell subpopulations and of their functions by cyto-chemical and cultural methods; (2) mineralogical studies by optical and electron microscopy to quantify the exposure; (3) biochemical studies (serum proteins, collagen precursors). Data on ambient exposure is collected, whenever possible.
The results of this investigation might help towards understanding the early reactions to inhalation of fibers and to foresee, within certain limits, the possible evolution into an overt lung disease. Moreover they will form the basis for further studies aimed at evaluating the possible effects of inhalation of known low concentrations of mineral fibers.

Progress : start = 01/83   end = 1987   publication =

Person(s) responsible : Prof. Gerolamo Chiappino, prof. Alessandra Forni.

Sponsor(s) : Center funds (Common budget for all scientific activities from various sources: Ministry of Education, Ministry of Labour, Industry)

Cooperation : Medizinisches Institut für Umwelthygiene an der Universität Düsseldorf (FRG).

Publication(s) on the project :
Chiappino G., Guerreri M.C., Forni A.: Bronchoalveolar lavage, a new tool in the diagnosis of occupational lung diseases. In: Abstracts XXI. Congr. Occup. Health, September 1984, Dublin, p. 226.
Forni A., Guerreri M.C., Chiappino G.: Nuovi metodi di indagine nelle pneumopatie professionali: il lavaggio broncoalveolare. Med. Lavoro 76, 11-16, 1985.

Codes : A.9.b) - A.11 - A.12

CENTRO STUDI E RICERCHE SUGLI EFFETTI BIOLOGICI DELLE POLVERI
INALATE, ISTITUTO DI MEDICINA DEL LAVORO, UNIVERSITA DI MILANO. (2)

Original language : Italian

Title : **CHANGES IN HAEMODYNAMICS OF PULMONARY CIRCULATION IN EARLY PNEUMOCONIOSIS (ESPECIALLY ASBESTOSIS).**

Research protocol :

As part of a larger investigation, started several years ago, on the changes in the haemodynamics of pulmonary circulation in occupational lung fibrosis (especially asbestosis), selected subjects with a history of exposure to asbestos with or without radiological evidence of asbestosis, are studied regarding the haemodynamics of pulmonary circulation using heart catheterization at rest and during work, and M-mode echocardiography. Cardiac index, mean pulmonary arterial pressure, total pulmonary vascular resistence and thickness of the free right ventricular wall are determined. In selected cases, the effect of vasodilators on pulmonary haemodynamics is studied in order to evaluate the reversibility of arterial pulmonary hypertension.

Progress : start = 1983   end = 1988   publication =

Person(s) responsible : Prof. Mario Tomasini.

Sponsor(s) : Center funds (Commun budget for all scientific activities from various sources: Ministry of Education, Ministry of Labour, Industry).

Cooperation : Servizio di Cardiologia, Istituti Clinici di Perfezionamento.

Publication(s) on the project :
Tomasini M., Villa A., Aresini G., Chiappino G. et al.: Studio emodinamico della circolazione polmonare in asbestosici con diverso grado di evidenza radiologica della pneumoconiosi. Med. Lavoro 71, 513-521, 1980.
Tomasini M., Lo Cicero G.: Effetti precoci delle lesioni polmonari da amianto sull'emodinamica del circolo polmonare. Osservazioni su 15 casi. Med. Lavoro 75, 306-312, 1984.

Codes : A.9.b) - A.11

Italy

CENTRO STUDI E RICERCHE SUGLI EFFETTI BIOLOGICI DELLE POLVERI
INALATE, ISTITUTO DI MEDICINA DEL LAVORO, UNIVERSITA DI MILANO. (3)

Original language : Italian

Title : **CLINICAL INVESTIGATION ON PLEURO-PULMONARY CHANGES IN SUBJECTS
EXPOSED TO MAN-MADE MINERAL FIBERS (MMMF).**

Research protocol :

A clinical and radiological study is being performed on selected subjects
with prolonged exposure to MMMF, in whom past or concomitant occupational
or extra-occupational exposure to asbestos can be ruled out, in order to
detect possible pleural and parenchymal lung changes.

Preliminary results on a small number of subjects exposed to glass fibers
are in print (see below): pleural plaques and slight parenchymal fibrotic
changes have been observed.

Progress : start = 1985   end = 1988   publication =

Person(s) responsible : prof. G. Chiappino, prof. M. Tomasini.

Sponsor(s) : Center funds (Common budget for all scientific activities from
various sources: Ministry of Education, Ministry of Labour, Industry).

Cooperation :

Publication(s) on the project : Tomasini M., Rivolta G., Chiappino G.: Effetti
sclerogeni attribuibili alla esposizione professionale a fibre vetrose in
un gruppo selezionato di lavoratori. Med. Lavoro, 1986 (in press).

Codes : A.9.b) - A.11

CENTRO STUDI E RICERCHE SUGLI EFFETTI BIOLOGICI DELLE POLVERI
INALATE, ISTITUTO DI MEDICINA DEL LAVORO, UNIVERSITA DI MILANO. (4)

Original language : Italian

Title : **FIBER CONCENTRATION IN LUNG TISSUE OF SUBJECTS NOT OCCUPATIONALLY EXPOSED TO ASBESTOS.**

Research protocol :

The project is aimed at clarifying the possible role of environmental
exposure to mineral fibers in the causation of lung cancer and
mesothelioma. For this purpose, the fiber concentration is being
determined in 100 consecutive apparently normal lung tissue samples
obtained from members of the general population undergoing surgery for
neoplastic or non-neoplastic thoracic disease. Part of the lung sample is
studied histologically. The subjects are carefully interviewed for place
of birth, place(s) of residence, occupation(s) and living habits, by an
independent interviewer. The study is carried out blindly. At the end of
the study the results obtained in subjects with neoplastic and
non-neoplastic thoracic disease will be compared and evaluated on the
basis of the anamnesic data.

Progress : start = 07/85   end = 12/88   publication =

Person(s) responsible : Prof. Gerolamo Chiappino, dott. Aldo Todaro.

Sponsor(s) : Center funds (Common budget for all scientific activities from
    various sources: Ministry of Education, Ministry of Labour, Industry).

Cooperation : Institut für Umwelthygiene an der Universität Düsseldorf (FRG) -
    Istituto di Clinica Chirurgica Generale e Terapia Chirurgica IV -
    Università di Milano.

Publication(s) on the project :

Codes : A.5 - A.10 - A.12

CENTRO DI STUDIO PER I PROBLEMI MINERARI,
c/o DIPARTIMENTO GEORISORSE E TERRITORIO—POLITECNICO.

Original language : Italian

Title : **INDUSTRIAL WASTES AND POLLUTION CAUSED BY ASBESTOS FIBRES.**

Research protocol :

A recent restriction on the unloading of industrial wastes containing asbestos is the starting point for a critical analysis of the definition of "free asbestos fibres" and for some suggestions about the methodology for the dosage of potentially dangerous fibres in these wastes. In particular, the author rules out the possibility of direct determination of asbestos fibres from the raw materials and suggests the separation of the fine fraction of these materials, and a subsequent analytical determination of the "free fibres".

Progress : start = 01/03/1985   end = 31/10/1985   publication =

Person(s) responsible : Enea Occella, professor of Mineral Dressing in the
                        Turin Polytechnic School.

Sponsor(s) : None.

Cooperation :

Publication(s) on the project :

Codes : A.5 - A.12 - C.0

ISTITUTO SUPERIORE DI SANITA-LABORATORIO DI ULTRASTRUTTURE.

Original language : Italian

Title : **MINERAL FIBRE SUBPROJECT: EVALUATION OF CANCER RISK FROM OCCUPATIONAL AND NON OCCUPATIONAL EXPOSURE TO AIRBORNE MINERAL FIBRES AND DUSTS**

Research protocol :

Objective:

The aim of the research project is the evaluation in Italy of the risk related to airborne fibers and dusts, by means of the study of the inorganic particulate in autoptic lung tissue samples from relevant sections of the population. Firstly, subjects who died of lung and bronchogenic carcinoma, mesothelioma, etc., will be studied. Then the investigation will be extended to groups of the population selected according to age, employment, place of residence, etc.

Aporoach:

After the mineralization of lung tissue samples, the resulting mineral fraction will be recovered and studied by high resolution electron microscopy, electron diffraction and X-ray microanalysis techniques.
The analytical data collected will be correlated to the statistical and anamnestic data relative to the groups under investigation.

Progress : start = 01/01/84   end = 31/12/88   publication =

Person(s) responsible : Prof. G. Donelli, Dr. L. Paoletti, Dr. F. Malchiodi Albedi, Dr. M.G. Petrelli.

Sponsor(s) : Istituto Superiore di Sanità "Progetto Ambiente", Consiglio Nazionale delle Ricerche - Progetto Finalizzato "Oncologia" and Progetto Finalizzato "Energetica".

Cooperation : See p. 194.

Publication(s) on the project : Donelli G. and Paoletti L. "Riconoscimento di fibre minerali mediante microscopia elettronica" In: Mesotelioma Maligno, Panel Nazionale dei Mesoteliomi and Gruppo die studio della SIPDTT eds., 1985, pp. 94-102.

Codes : A.0 - A.9.b) - A.10 - A.11 - A.12 - B.0 - B.3 - B.5 - B.6 - B.9 - B.10 - C.0

ITALIAN RAILWAY HEALTH SERVICE.

Original language : Italian

Title : **HIGH SECURITY DEPARTMENT FOR ELIMINATING ASBESTOS USED IN AN ITALIAN RAILWAY FACTORY**

Research protocol :

Until 1975, asbestos was used in the insulation of railway cars. In order to avoid health dangers for workers repairing and servicing these cars, Italian railways have started a plan for eliminating asbestos from old cars.

A high security department built for the purpose is described in the study. Workers' health is protected by the strict separation of the different zones of the department and by individual means of protection.

In order to protect the working environment, waste material is collected immediately and conveyed to controlled dumps. Air and water purifications is also provided.

Progress : start = 01/10/82   end = 15/6/85   publication =

Person(s) responsible : C. Mingozzi, P. Pieri, A. Serio, A. Vitale.

Sponsor(s) : Italian Railways.

Cooperation :

Publication(s) on the project : Proceeding of the 48th Congress of the Italian Association of Occupational Health and Industrial Hygiene - September 1985.

Codes : A.11 - A.12 - B.4 - B.5

SERVIZIO DI ANATOMIA E ISTOLOGIA PATOLOGICA, OSPEDALE DI MONFALCONE.

Original language : Italian

Title : **ASBESTOS EXPOSURE IN THE MONFALCONE AREA, ITALY.**

Research protocol :

The Monfalcone area, in north-eastern Italy, is a small industrial region (60,000 inhabitants), whose main industry is shipbuilding. Three parameters of asbestos exposure (hyalin plaques of the pleura, lung asbestos bodies, and work history) were investigated in a series of 1049 consecutive necropsies, carried out at the Monfalcone Hospital between October 1979 and December 1984. Lifetime occupational histories were obtained from the patients'relatives by personal interviews. Routine lung sections were examined for asbestos bodies in all cases; in addition isolation and quantification of lung asbestos bodies were performed in one third of the necropsies (Smith and Naylor method). The relationships between the above parameters and various pathological conditions (malignancies, liver cirrhosis, Alzheimer's disease, etc.) are analyzed.

Progress : start = 1979   end = 1985   publication = 1986

Person(s) responsible : Claudio Bianchi MD

Sponsor(s) : Unità Sanitaria Locale n° 2 - Goriziana.

Cooperation :

Publication(s) on the project : Proceedings of the International Congress on "Risk assessment of occupational exposures in the harbour environment" (Genova, Italy, 3-5 Oct. 1984).

Codes : A.10 - B.6 - C.0

87

Japan

ASAHI GLASS CO., LTD., RESEARCH LABORATORY.

Original language : English

Title : **NON-ASBESTOS SHEETS BY HATSCHEK PROCESS**

Research protocol :

    The development of non-asbestos cement sheets by means of the Hatschek process has been carried out for about five years. Glassfibers and other sorts of fibers were used instead of asbestos. The non-asbestos cement sheet could be manufactured by the appropriate combination of these fibers and mix proportions of matrix. Suitable mix proportions for the Hatschek process were selected by a filtration test. They were first applied to a small model of the Hatschek machine, which was useful for studying whether the mix proportion was suitable. It was finally confirmed that it was possible to manufacture non-asbestos cement sheets using a commercial plant. The bending strength was 23 MN/m². The density was 1.53 g/cm³. The sheet has good incombustibility and durability. It is also easy to cut and drill the sheet. It can be used for exterior walls, interior walls, partitions, ceilings, etc.

Progress : start =   end =   publication =

Person(s) responsible : T. Ohigashi.

Sponsor(s) :

Cooperation :

Publication(s) on the project :

Codes : A.12 - A.14 - B.4 - C.4 - C.7 - D.1 - D.3.a) - E.11 - E.37

DEPARTMENT OF COMMUNITY MEDICINE, TOYAMA MEDICAL AND PHARMACEUTICAL UNIVERSITY.

Original language : Japanese

Title : **MONITORING OF CYTOGENETIC HEALTH EFFECTS ON PERSONS WITH A HISTORY OF ASBESTOS EXPOSURE AND THEIR FAMILIES**

Research protocol :

A registration system of persons with a history of asbestos exposure, by means of compulsory health examination for workers involved in the asbestos industries, is being established in our district. The cytogenetic monitoring of persons associated with asbestos exposure and their families has been carried out by the observation of sister chromatid exchange (SCE) and micronucleus in cultured peripheral blood lymphocytes.

Furthermore, their sensitivity to mitomycin C, which is able to enhance the development of SCE and micronucleus, was also tested.

Among those persons currently exposed to asbestos and organic solvents, certains abnormalities were found.

Progress : start = 01/11/84   end = 01/11/89   publication = 01/11/90

Person(s) responsible : Sadanobu Kagamimori, MD.

Sponsor(s) : Local Government.

Cooperation : Local Government, the Public Health Laboratory Association.

Publication(s) on the project :

Codes : A.9.a) – A.9.b) – A.10

DEPARTMENT OF HYGIENE, KAWASAKI MEDICAL SCHOOL.

Original language : English

Title : **PROLIFERATION STIMULATING EFFECT OF ASBESTOS FIBERS ON HUMAN
B-LYMPHOCYTE CELL LINE: IN SERUM FREE MEDIUM**

Research protocol :

Human B-lymphocyte cell lines, SB and Raji cells, were incubated with
asbestos fibers, chrysotile, crocidolite, amosite & anthophyllite, in
serum free medium HB 101 (Hana Media, Inc. USA). After 48 hrs the rate of
DNA synthesis was measured with H3-thymidine using liquid scintillation
system (Aloka, LSC-900). As reported previously, in the case of the medium
supplemented with fetal calf serum (Clin. exp. Immunol, 56: 425-430,
1984), Raji cells were stimulated with asbestos fibers, especially
crocidolite and amosite, in the concentration of 20 μg/ml in the serum
free medium. Anthophyllite stimulated Raji cells only slightly, and
chrysotile was cytotoxic to Raji cells.

Crocidolite stimulated slightly SB cells, and amosite or chrysotile were
cytotoxic to SB cells.

Progress : start = 01/04/84   end = 31/03/86   publication =

Person(s) responsible : Ayako Ueki.

Sponsor(s) : Kawasaki Medical School, Grant No. 58-803.

Cooperation :

Publication(s) on the project :

Codes : A.9.a) – A.11 – A.12 – B.0 – C.0

Poland

ACADEMY OF MINING AND METALLURGY, INTERBRANCH
INSTITUTE OF BUILDING AND REFRACTORY MATERIALS.

Original language : Polish

Title : **ASBESTOS GRINDING IN A ROD MILL**

Research protocol :

The grinding of asbestos takes place in a rod mill.

The low strength rods simultaneously press on a greater number of asbestos
fibres than in a chaser mill (on a bigger surface). Asbestos was put into
the rod mill as a water asbestos suspension. Its concentration, the
rotation of the mill, the number and diameter of the rods are parameters
which regulate the grinding process. These parameters are the subject of
our investigations.

Progress : start = 01/01/81   end =   publication =

Person(s) responsible : D. Se. Eng. Jerzy R. L. Dyczek.

Sponsor(s) : Ministry of Education and Ministry of Building.

Cooperation :

Publication(s) on the project : Asbestos   grinding   in   a   rod   mill   (under
        preparation).

Codes : A.0 - A.5 - C.1 - C.2 - D.1 - D.3

South Africa

DEPARTMENT OF MINERAL AND ENERGY AFFAIRS, AIR QUALITY RESEARCH.

Original language : English

Title : **ENVIRONMENTAL SAMPLING FOR RESPIRABLE ASBESTOS FIBRES IN ASBESTOS MINING FIELDS IN THE REPUBLIC OF SOUTH AFRICA**

Research protocol :

   Monitoring of environmental samples for respirable asbestos fibres using the Scanning Electron Microscope as prescribed by the Recommended Technical Method No. 2 (RTM 2) as devised by the Asbestos International Association.

Progress : start = 13/01/83   end = 31/12/86   publication = 12/87

Person(s) responsible : Mr. J.H.O. van Sittert.

Sponsor(s) : Department of Mineral and Energy Affairs.

Cooperation : Griqualand Exploration and Finance Co.

Publication(s) on the project :

Codes : A.2 - A.12

CLINIC OF OCCUPATIONAL MEDICINE.

Original language : Swedish

Title : **LUNG CANCER AMONG WORKERS EXPOSED TO MAN-MADE MINERAL FIBRES (MMMF).**

Research protocol :

Man-made mineral fibres (MMMF) have been used for the insulation of wooden houses in Sweden since the 1940's. Both rockwool and glasswool have been used. This is a cohort study of MMMF-exposed workers employed in the production of wooden houses.

Around 1500 workers are included in the cohort. Lung cancer incidence and mortality are compared to expected numbers computed from national rates and standardized for the geographic region.

The current levels of exposures to MMMF are measured in all the factories included. Earlier production conditions will be reconstructed to enable an estimation of past exposure.

Progress : start = 1986   end = 1988   publication =

Person(s) responsible : Doctor Per Gustavsson, Professor Christer Hogstedt.

Sponsor(s) :

Cooperation :

Publication(s) on the project :

Codes : A.10 - A.11 - D.3 - E.11 - E.37 (MMMF).

DEPARTMENT OF ONCOLOGY, UNIVERSITY HOSPITAL UMEA.

Original language : Swedish

Title : **A CASE CONTROL STUDY ON COLONCANCER WITH SPECIAL REGARD TO ASBESTOS EXPOSURE**

Research protocol :

The cases in the study consist of all living patients less than 75 years of age with colon cancer, who were diagnosed 1980-83 and reported to the Swedish Cancer Registry. They are all residents of the catchment area of the Department of Oncology in Umeå, i.e the counties of Norrbotten, Västerbotten and Västernorrland. Since reporting to the Swedish Cancer Registry is compulsory, almost all cancer cases can be identified through this register.

Approximately 300 cases are included in the study. For each case, 2 controls are drawn from the National Population Register and matched for sex, age and county. Information about various exposures is obtained by written questionnaires consisting of 16 pages with various questions about previous jobs, different types of chemical exposure during employ- ment or leisure time, intake of coffee, tea and alcohol, food habits, smoking habits, previous diseases, intake of drugs, etc. If necessary, questionnaires are completed over the phone by an interviewer.

The study will be completed some time in 1986.

Progress : start = 1983   end = 1986   publication =

Person(s) responsible : Nils-Olof Bengtsson MD, Lennart Hardell MD, Olav Axelsson Prof.

Sponsor(s) : Swedish Cancer Society.

Cooperation :

Publication(s) on the project :

Codes : A.10 - B.14

NATIONAL BOARD OF OCCUPATIONAL SAFETY AND HEALTH. (1)

Original language : Swedish

Title : **DEVELOPMENT OF A METHOD FOR DETERMINING THE ASBESTOS CONTENT IN BULK SAMPLES**

Research protocol :

In Sweden, bulk samples with more than 1% asbestos are classified as asbestos containing materials and special precautions must be observed. No reliable methods have been available for the determination of small amounts of asbestos in such samples. The aim of this study is to investigate the possibility of developing a reliable and accurate method for the qualitative and quantitative determination of asbestos in bulk samples. X-ray diffractometry is to be used as the analytical tool.

Progress : start = 01/07/82   end = 30/06/86   publication =

Person(s) responsible : Lennart Lundgren, Staffan Krantz.

Sponsor(s) :

Cooperation :

Publication(s) on the project :

Codes : A.12 - B.3 - C.0

Sweden

NATIONAL BOARD OF OCCUPATIONAL SAFETY AND HEALTH. (2)

Original language : Swedish

Title : **IDENTIFICATION AND QUANTIFICATION OF AIRBORNE INORGANIC FIBRES USING SCANNING ELECTRON MICROSCOPY.**

Research protocol :

The aim is to develop an in-house method for identification and quantification of airborne fibres sampled on membrane filters. The following parameters are to be investigated: sampling filters, sample preparation, asbestos identification criteria, fields of application, counting statistics and detection levels.

Progress : start = 01/01/84   end = 31/12/85   publication = 01/06/85

Person(s) responsible : Staffan Krantz, Birgit Paulsson.

Sponsor(s) : Work Environmental Fund.

Cooperation :

Publication(s) on the project :

Codes : A.12 - B.3 - C.0

NATIONAL BOARD OF OCCUPATIONAL SAFETY AND HEALTH. (3)

Original language : Swedish

Title : **DIRECT IDENTIFICATION OF AIRBORNE FIBRES ON MEMBRANE FILTERS.**

Research protocol :

    The project investigates the use of Quantitative Interference Contrast Microscopy for the analysis of airborne fibres collected on membrane filters. This microscopy technique makes it possible to determine the refractive index (R.I.) of a fiber in a known filter matrix.

    The method developed has been used for the determination of R.I. of fibres collected on membrane filters at different workplaces. Asbestos fibres as well as Man Made Mineral Fibres and ceramic fibres have been included in the study.

Progress : start = 01/01/84   end = 31/12/85   publication =

Person(s) responsible : Lennart Lundgren, Gunbritt Berglund.

Sponsor(s) : National Board of Occupational Safety and Health.

Cooperation :

Publication(s) on the project :

Codes : A.12 - B.3 - C.0

UNIVERSITY OF UPPSALA, DEPARTMENT OF LUNG MEDICINE.

Original language : English

Title : **ASBESTOS-RELATED PLEUROPULMONARY CHANGES**

Research protocol :

All persons in the county of Uppsala, Sweden, who are found to have chest X-ray changes are registered and given check-ups at least every second year. Other investigations (lung function, computerized tomography, broncoscopy, etc) are made if clinically important and as part of smaller studies. To date, more than 1500 persons have been included, and important information about incidence, prognosis, complications etc. has been gathered.

The cohort is followed prospectively and a follow-up of the findings so far is now in preparation and we hope to have it published in 1986.

Progress : start = 1975   end = 2000   publication =

Person(s) responsible : Gunnar Hillerdal, MD,
                        Ass. Prof., Dept. of Lung Medicine

Sponsor(s) :

Cooperation :

Publication(s) on the project : About a dozen published small reports so far.

Codes : A.11 - B.6

INSTITUT UNIVERSITAIRE DE MEDECINE DU TRAVAIL ET D'HYGIENE INDUSTRIELLE.

Original language : French

Title : **NEW STRATEGIES FOR THE ASSESSMENT OF ASBESTOS EXPOSURE IN PARAOCCUPATIONAL AND OCCUPATIONAL ENVIRONMENTS**

Research protocol :

- A new strategy was developed to assess the exposure to asbestos fibers in paraoccupational indoor environments. Fibrous Aerosol Monitor and Phase Contrast Microscopy are used as screening methods, with reference to Transmission Electron Microscopy for absolute results. A number of asbestos insulated buildings were studied, with particular attention to the influence of human activity on the airborne fiber level.

- The urine of two groups of people (one occupationally exposed group and one control group) was analysed to determine the asbestos fiber concentration using Transmission Electron Microscopy, with the objective of obtaining a biological index of exposure to airborne asbestos.

Progress : start = 01/10/83   end = 30/09/85   publication = 1986

Person(s) responsible : Prof. M. Guillemin.

Sponsor(s) : Swiss National Research Fund.

Cooperation : Institute of Electron Microscopy (Dr. Ph. Buffat) and Depart-
    ment of Architecture (F. Iselin), Swiss Federal Institute of Technology
    at Lausanne.

Publication(s) on the project : Influence of the human activity on the
    airborne fiber level in paraoccupational environments, J. Air Poll.
    Contr. Assoc. <u>35</u>, 836 (1985).

Codes : A.7 - A.12

SERVICE CANTONAL D'ECOTOXICOLOGIE, GENEVE.

Original language : English

Title : **PROBLEMS OF DIFFERENTIATION BETWEEN ASBESTOS AND KAOLINITE IN COMPOSITE MATERIALS**

Research protocol :

In screening various building materials for asbestos, it appears that attempts to differentiate between asbestos and kaolinite by the usual optical or even X-ray diffraction methods lead to ambiguous inter-pretations.

Various analytical approaches such as sample preparation and infra-red spectroscopy are studied in an effort to solve this questions.

Progress : start = 09/85   end =   publication =

Person(s) responsible : two.

Sponsor(s) :

Cooperation : Département de Minéralogie, Museum de Genève.

Publication(s) on the project :

Codes : A.5 - B.9 - C.1 - C.7

United Kingdom

ANJALENA PUBLICATIONS LTD. (1)

Original language : English

Title : **SCIENTIFIC ADVANCES IN ASBESTOS 1967 TO 1985.**

Research protocol :

A detailed review of scientific research and development on asbestos which spans the occasions of four international asbestos conferences in 1967, 1971, 1975 and 1980. The conference papers are available only through a limited printing of proceedings, and it is the purpose of this review to link these with the considerable amount of other information published during the period. The bibliography covers 400 references.

This project has been started in response to numerous requests from scientific, industrial and medical authorities for a consolidated and detailed sourcebook on the properties of asbestos beyond what is available in other textbooks.

Progress : start = 06/82   end = 12/85   publication = 03/86

Person(s) responsible : A.A. Hodgson.

Sponsor(s) : Anjalena Publications Ltd.

Cooperation : Asbestos International Association.

Publication(s) on the project :

Codes : A.1 - A.2 - A.3 - A.4 - A.5 - A.12 - B.0 - C.1 - C.2

United Kingdom

ANJALENA PUBLICATIONS LTD. (2)

Original language : English

Title : **ALTERNATIVES TO ASBESTOS AND ASBESTOS PRODUCTS, 2nd Edition.**

Research protocol :

This is an ongoing account of progress made in the search for and use of alternatives to asbestos. It does not seek to promote alternatives, but examines the properties, costs and availabilities of the likely and unlikely raw materials which are being used or promoted. It assesses the nature and characteristics of alternative non-asbestos products in the six major product groups of fibre-reinforced cements; building materials and insulations; heat resistant textiles, friction products, reinforced plastics; and packings and jointings.

Progress : start = 06/85   end = 12/86   publication = 03/87

Person(s) responsible : A.A. Hodgson.

Sponsor(s) : Anjalena Publications Ltd.

Cooperation : Asbestos International Association.
                Asbestos Information Centre, U.K.

Publication(s) on the project : Alternatives   to   Asbestos   and   Asbestos
        Products, 1st. Edition, January, 1985.

Codes : D.1 - D.2 - D.3 - D.4 - D.5 - D.6 - D.7 - D.8 - D.9 - D.10 - D.11 -
        D.12 - E.2 - E.3 - E.4 - E.5 - E.6 - E.11 - E.12 - E.13 - E.15 - E.16 -
        E.18 - E.19 - E.20 - E.21 - E.22 - E.23 - E.24 - E.25 - E.26 - E.27 -
        E.28 - E.29 - E.30 - E.31 - E.32 - E.33 - E.35

United Kingdom

CONSTRUCTION MATERIALS RESEARCH GROUP,
CIVIL ENGINEERING DEPARTMENT, UNIVERSITY OF SURREY.

Original language : English

Title : **ALTERNATIVES TO ASBESTOS CEMENT BASED ON CONTINUOUS POLYMER NETWORKS.**

Research protocol :

A major research effort since 1976 has been the development of a patented new material for thin sheet roofing and cladding as an alternative to asbestos-cement. This development has involved the use of opened layers of polypropylene networks in a cement matrix and the reinforcing mechanism and the main material characteristics have been established during this period. Major industrial resources have been devoted to the development of production machinery both for the networks and for the composite which is known by the Trade Name NETCEM. Production licences have been sold throughout Europe and Australia.

Financial support has been provided by the University of Surrey, U.K. Governement sources & licences. Major industrial developments have been provided by Italian licensees with whom extensive technical co-operation has been made since 1978.

Progress : start = 1978   end = 1980   publication = continuing

Person(s) responsible : Dr. D.J. Hannant, Mr. J.J. Zonsveld.

Sponsor(s) : University of Surrey.

Cooperation : Major Italian industrial concerns.

Publication(s) on the project : Hannant D.J. and Zonsveld J.J., "Polyefin fibrous networks in cement matrices for low cost sheeting", Philosophical Transactions of Royal Society, 1980, A.294, pp. 591-597.
Keer J.G. and Thorne A.M, "Performance of polypropylene-reinforced cement sheeting elements", International Symposium on Fibre-reinforced Concrete, Detroit, USA, September 1982.
Hibbert A.P. and Hannant D.J., "Toughness of cement composites containing polypropylene films compared with other fibre cements". Composites, October 1982, pp. 393-399.

Codes : A.0 - A.8 - B.0 - C.5.b) - D.1 - E.37

FORMER AEROSOL LABORATORY, UNIVERSITY OF ESSEX.

Original language : English

Title : **A PERSONAL SAMPLER FOR AIRBORNE ASBESTOS DUST**

Research protocol :

Asbestos diseases and health standards are reviewed. Asbestosis is caused by breathing fibres longer than 10-15 µm, shorter fibres being cleared from the lungs by phagocytes. Soft chrysotile is much more active a cause of asbestosis than other varieties. Lung carcinoma is also caused by soft fibres and is peribronchiolar in origin because the curled, splayed fibres do not work their way through the lungs due to respiratory movements. Hard, needle shaped, smooth fibres of the amphiboles travel through lung tissue and, if long enough to escape phagocytosis, cause carcinoma. Mesothelioma of the pleura is caused by long inhaled fibres of amphibole which have passed from airways to the pleura; it can be excited by injection or application to the pleura of animals of chrysotile fibres of any length but does not result from the inhalation of chrysotile fibres which anchor themselves in lung tissue and cannot travel to the pleura. Selective sampling of airborne asbestos dust is necessary to pick out the small proportion of disease causing fibres from the large amount of short-fibred material which is disposed of by phagocytes. The selective sampler must retain fibres longer than 10 to 15 µm with diameters from 0.1 to 0.5 µm; a basic design for such a sampler is described.

Progress : start = 1982   end = 1983   publication =

Person(s) responsible : C.N. Davies.

Sponsor(s) : Health & Safety Executive, Steel City House, West Street Sheffield.

Cooperation :

Publication(s) on the project :

Codes : A.0 - B.3 - C.1 - C.2

HEALTH & SAFETY EXECUTIVE, EPIDEMIOLOGY AND MEDICAL STATISTICS UNIT.

Original language : English

Title : **ASBESTOS WORKERS SURVEY.**

Research protocol :

Maintenance of a register of workers exposed to asbestos that is linked to a national death registration system to monitor mortality in different time periods and industries. As newly exposed workers are identified they are included on the register. Data on smoking habit is recorded. There is a possibility of some work being undertaken to assess the links between dust exposure levels and mortality rates in different industry sectors.

Progress : start = 1970   end = continuing   publication =

Person(s) responsible : Dr. R.D. Jones, Mr J.T. Hodgson.

Sponsor(s) : Health & Safety Executive.

Cooperation : Office of Population Censuses and Surveys.

Publication(s) on the project : One report to date which will shortly be published in the British Journal of Industrial Medicine – Mortality of Asbestos Workers in England and Wales 1971-81.

Codes : A.10 – A.11 – C.0 – D.0

United Kingdom

HEALTH & SAFETY EXECUTIVE, RESEARCH AND LABORATORY SERVICES DIVISION.

Original language : English

Title : **DIRECT READING FIBRE MONITOR**

Research protocol :

There is a need for a direct reading fibre monitor for use in asbestos manufacturing plants and in asbestos stripping operations.

The instrument would not replace the optical microscope method of fibre counting, but would act as indication of things going wrong.

For the instrument to be of any practical use it must in general follow optical fibre counts, and be cheap to purchase.

Progress : start = 01/01/84   end = 01/01/87   publication =

Person(s) responsible : Dr. A.P. Rood & Mr. E.J. Walker

Sponsor(s) : Health & Safety Executive.

Cooperation :

Publication(s) on the project : Internal report: IR/L/DI/84/5.

Codes : A.4 - A.7 - A.12 - B.3 - C.0

INSTITUTE OF OCCUPATIONAL MEDICINE. (1)

Original language : English

Title : **THE PATHOGENIC POTENTIAL OF SHORT FIBRE CHRYSOTILE ASBESTOS.**

Research protocol :

Long term inhalation and injection studies in rats using a short fibre chrysotile preparation and for comparison purposes a chrysotile preparation with the highest possible proportion of long fibres. Animals exposed by inhalation to dust cloud of 10 mg/m^3 for 12 months with full life span follow-up.

In injection studies animals received a single intraperitoneal injection of either 25 mg, 2.5 mg or 0.25 mg of dust.

Progress : start = 01/01/83   end = 31/12/86   publication = 06/87

Person(s) responsible : Dr. J.M.G. Davis.

Sponsor(s) : Asbestos Institute (Institut de l'Amiante).

Cooperation :

Publication(s) on the project :

Codes : A.9.a) - B.7 - C.1

United Kingdom

INSTITUTE OF OCCUPATIONAL MEDICINE. (2)

Original language : English

Title : **STUDIES ON THE EFFECTS OF ELECTROSTATIC CHARGE ON THE PATHOGENIC.**

Research protocol :

Long term inhalation studies in rats using preparations of UICC Rhodesian chrysotile with either normal electrostatic charge or after discharging by radioactive source.

Dose level 10 mg/m^3 - for 12 months with full life span follow-up.

Progress : start = 30/06/82   end = 30/06/85   publication = 06/86

Person(s) responsible : Dr. J.M.G. Davis.

Sponsor(s) : British Asbestosis Research Council.

Cooperation :

Publication(s) on the project :

Codes : A.9.a) - B.7 - C.1

INSTITUTE OF OCCUPATIONAL MEDICINE. (3)

Original language : English

Title : **STUDIES ON THE EFFECTS ON THE PATHOGENICITY OF CHRYSOTILE AND AMOSITE ASBESTOS OF INERT DUST (TITANIUM DIOXIDE) INHALED AT THE SAME TIME.**

Research protocol :

Long term inhalation studies in rats. Animals exposed to either chrysotile or amosite (dusts prepared from raw commercial long fibre grades) at 10 mg/m^3 plus titanium dioxide at 10 mg/m^3. Exposure period 12 months with full life span follow-up.

Progress : start = 01/03/85   end = 01/03/88   publication = 09/88

Person(s) responsible : Dr. J.M.G. Davis.

Sponsor(s) : British Asbestosis Research Council.

Cooperation :

Publication(s) on the project :

Codes : A.9.a) — B.7 — C.1 — C.2

United Kingdom

INSTITUTE OF OCCUPATIONAL MEDICINE. (4)

Original language : English

Title : **DOSE RESPONSE STUDIES OF MESOTHELIOMA PRODUCTION WITH ERIONITE.**

Research protocol :

Intraperitoneal injection studies in rats. Animals given single injection of dust at dose levels ranging from 25 mg down to 0.005 mg. Both numbers of mesotheliomas will be recorded and the tumour induction period.

Progress : start = 01/01/85   end = 31/12/87   publication = 06/88

Person(s) responsible : Dr. R.E. Bolton.

Sponsor(s) : British Asbestosis Research Council.

Cooperation :

Publication(s) on the project :

Codes : A.9.a) - B.7 - C.5.a)

INSTITUTE OF OCCUPATIONAL MEDICINE. (5)

Original language : English

Title : **THE EFFECTS OF MULTIPLE DOSING ON MESOTHELIOMA PRODUCTION
WITH CHRYSOTILE ASBESTOS.**

Research protocol :

   Intraperitoneal injection studies in rats. Animals given either a single
   intraperitoneal injection of chrysotile or the same mass dose given as a
   series of injections at various time points.

Progress : start = 01/06/83   end = 31/05/86   publication = 12/86

Person(s) responsible : Dr. R.E. Bolton.

Sponsor(s) : British Asbestosis Research Council.

Cooperation :

Publication(s) on the project :

Codes : A.9.a) - B.7 - C.1

United Kingdom

INSTITUTE OF OCCUPATIONAL MEDICINE. (6)

Original language : English

Title : **CELL RECRUITMENT STUDIES IN RAT LUNG TISSUE FOLLOWING THE INHALATION OF ASBESTOS.**

Research protocol :

At various time points during and after short periods (6 weeks) of exposure to either chrysotile or amosite asbestos, rats are killed and the lungs subjected to pulmonary lavage to examine the populations of free cells that can be obtained. Studies on these cells include differential counts, chemotactic ability and production of bioactive materials such as enzymes and reactive oxygen intermediates.

Progress : start = 01/01/84   end = 31/12/87   publication = 06/88

Person(s) responsible : Dr. K. Donaldson.

Sponsor(s) : British Asbestosis Research Council.

Cooperation :

Publication(s) on the project :

Codes : A.9.a) – B.7 – C.1 – C.2

INSTITUTE OF OCCUPATIONAL MEDICINE. (7)

Original language : English

Title : **STUDIES OF THE LUNG ASBESTOS CONTENT IN NON INDUSTRIALLY EXPOSED HUMANS.**

Research protocol :

A series of lungs from members of a British urban population are being examined for their asbestos content. A detailed questionnaire on work history is presented to next of kin to eliminate as far as possible those with industrial exposure to asbestos.

Progress : start = 01/01/85   end = 31/12/87   publication = 06/88

Person(s) responsible : Dr. V.A. Ruckley.

Sponsor(s) : British Asbestosis Reserch Council.

Cooperation :

Publication(s) on the project :

Codes : A.10 - A.12 - B.14 - C.0

INSTITUTE OF OCCUPATIONAL MEDICINE. (8)

Original language : English

Title : **INTERNATIONAL COUNTING TRIAL FOR ASBESTOS DUST EXTRACTED FROM HUMAN LUNGS.**

Research protocol :

The importance of the accurate estimation of lung asbestos content is increasing in the medico-legal field. However, it is known that different laboratories can produce markedly different figures from any one case.

The Institute of Occupational Medicine in Edinburgh is organising an international counting trial which will involve over 40 laboratories throughout the world. Specimens will be circulated and counts compared in order to examine counting differences and perhaps determine their cause.

Progress : start = 01/06/85   end = 31/05/88   publication = 12/88

Person(s) responsible : Dr. V.A. Ruckley.

Sponsor(s) : British Asbestosis Research Council.

Cooperation :

Publication(s) on the project :

Codes : A.11 - A.14 - B.6 - C.0

MRC ENVIRONMENTAL EPIDEMIOLOGY UNIT.

Original language : English

Title : **A FOLLOW-UP OF THE WORKFORCE OF A CHRYSOTILE ASBESTOS CEMENT FACTORY.**

Research protocol :

   A standard cohort mortality follow-up is being carried out on about
   2,500 workers in one factory in England. The National Health Service
   Central Register has been used to determine vital status and cause of
   death of these employees. At a pre-analysis stage individuals have been
   assigned to qualitative exposure levels based on their recorded job
   histories at the factory and extensive discussions about the characters
   of jobs with long termed employees of the firm. Comparisons will be made
   between mortality from various causes, including cancers, and the
   experience of appropriate populations of men of similar ages.

Progress : start = 01/01/84   end = 30/06/86   publication = 30/06/86

Person(s) responsible : Professor M.J. Gardner.

Sponsor(s) : Health and Safety Executive.

Cooperation :

Publication(s) on the project :

Codes : A.10 - A.11 - C.1 - D.1

United Kingdom

PATHOLOGY DEPARTMENT, MANCHESTER UNIVERSITY.

Original language : English

Title : **LUNG ASBESTOS FIBRE CONTENT OF ASBESTOS RELATED PULMONARY DISEASE.**

Research protocol :

The asbestos fibre content of the left upper and lower lobes of subjects who have been occupationally exposed to asbestos is estimated.

The fibres are extraced from formalin fixed lungs by potassium hydroxide digestion and the fibre count estimated by phase contrast microscopy.

The extent of asbestos exposure is, therefore, estimated by the numbers of fibres (coated and uncoated) found in the lung digest. The figures are expressed as number of fibres per gram of dried lung. The fibre count is related to the following asbestos-related conditions: hyaline pleural plaques, bronchial carcinoma, asbestosis and pleural mesothelioma.

Progress : start = 1981   end = ongoing   publication =

Person(s) responsible : Dr. A.W. Jones.

Sponsor(s) : None.

Cooperation : Manchester Pneumoconiosis Medical Panel.

Publication(s) on the project :

Codes : A.11 - B.2 - B.3 - B.6

RESPIRATORY INVESTIGATION CENTRE, BELFAST CITY HOSPITAL.

Original language : English

Title : **A FOLLOW-UP OF ASBESTOS-EXPOSED INSULATION WORKERS IN BELFAST**

Research protocol :

   A long-term follow-up study started in 1965 of all the Insulation
   Workers in Belfast who are exposed to asbestos.

Progress : start = 1965   end =   publication =

Person(s) responsible : Dr. Jean H.M. Langlands, M.D., F.R.C.P. Ed.

Sponsor(s) :

Cooperation :

Publication(s) on the project :
   Wallace WFM, Langlands JHM, Insulation workers in Belfast 1. Brit J
   Industr Med 1971; 28: 211-6. Langlands JHM, Wallace WPM, Simpson MJC.
   Insulation workers in Belfast 2. Brit J. indust Med 1971; 28: 217-25.
   Langlands JHM, Coyne FC. Simpson MJC, Campbell MJ. Follow-up of asbestos
   insulation workers in Belfast. Thorax 1981; 36: 232.
   Campbell MJ, Langlands JHM. Analysis of a follow-up study. An example
   from asbestos-exposed insulation workers. Scand j work environ health
   1982; 8: suppl 1, 43-7

Codes : A.10 - D.3

STRANGEWAYS RESEARCH LABORATORY.

Original language : English

Title : **AN INVESTIGATION INTO THE CARCINOGENIC ACTION OF METALLIC AND MINERAL PARTICULATES INTRODUCED INTO THE LUNG BY INTRATRACHEAL INSTILLATION.**

Research protocol :

Many kinds of industrial and environmental airborne particulates have been implicated in the causation of human lung cancer, and it is suspected that low levels of systemic carcinogens may act synergistically with these so that some exposed individuals are at increased risk.

The aim of this investigation is to quantify the carcinogenic activity, and other pathological effects, of very low levels of chrysotile asbestos administered alone and in combination with various particulate metals (Cd, Ni and Zn) and/or other putative carcinogens (lead oxide and benzo(a)pyrene), including systemic ones (caffeine and N-nitroso-heptamethyleneimine), using non-inhalation techniques in the rat. All particulate materials were introduced into the lung by intra-tracheal instillation. It is hoped particularly to identify combinations of agents which act synergistically with chrysotile in the causation of lung cancer.

Progress : start = 01/10/82   end = 30/09/86   publication = 30/12/86

Person(s) responsible : J.C. Heath & P.T.C. Harrison.

Sponsor(s) : Health & Safety Executive.

Cooperation : L.S. Levy, Institute of Occupational Health, University of Birmingham.

Publication(s) on the project :

Codes : A.9.a) - B.7 - C.1 - C.7

WATER RESEARCH CENTRE ENGINEERING.

Original language : English

Title : **USAGE, PERFORMANCE AND DEGRADATION OF ASBESTOS CEMENT WATER MAINS IN THE U.K.**

Research protocol :

> This study is to investigate the use of asbestos cement pressure pipe in the U.K. water supply system, including details of the lengths of AC in use and the diameter and age profiles of the system. The engineering performance of AC pipe is to be examined from analysis of failure statistics from selected areas. To develop an understanding of the factors controlling degradation of AC, an extensive pipe sampling exercise is to be performed with analysis of the exhumed pipes using various chemical and mechanical techniques. Various options for renovating deteriorated pipes are to be examined and their effect on levels of fibres in the conveyed water are to be monitored.

Progress : start = 01/04/84   end = 31/03/87   publication =

Person(s) responsible : E.P. White.

Sponsor(s) : Department of the Environment.

Cooperation : U.K. Water Utilities.

Publication(s) on the project :

Codes : A.0 - C.0 - D.1

U.S.A.

EAST BAY MUNICIPAL UTILITY DISTRICT, OAKLAND.

Original language : English

Title : **ASBESTOS REMOVAL IN CONVENTIONAL WATER TREATMENT PROCESSES.**

Research protocol :

East Bay Municipal Utility District (EBMUND) is undertaking a study to determine the extent to which asbestos is removed by filtration, coagulation/filtration, and coagulation/ flocculation/sedimentation/filtration. Data will be collected through June 1986. Preliminary results show that asbestos is nearly always removed to a greater extent than is turbidity; however, asbestos removal and turbidity removal are not proportional.

Progress : start = 20/04/84   end = 30/06/86   publication =

Person(s) responsible : M.L. Price.

Sponsor(s) : EBMUD.

Cooperation :

Publication(s) on the project :

Codes : A.12 - B.10

THE GEORGE WASHINGTON UNIVERSITY MEDICAL CENTER. (1)

Original language : English

Title : **CANCER IMMUNOPROPHYLAXIS**

Research protocol :

>
> Preliminary pilot work is published in:
> Cancer Detection and Prevention volume 3 no. 2: 419-448 (1981) and deals
> with the possibility of immunoprophylaxis in high-risk subjects exposed
> to asbestos and as heavy smokers and within a given age group.
> Another aspect is the necessity for monitoring of such a study and a new
> technique has been developed using genetic engineering with preliminary
> studies reported in Cancer Detec. Prev. 6: 185-191 (1983).
> These two aspects, the clinical protocol and the monitoring techniques,
> constitute the ongoing study.

Progress : start =   end =   publication =

Person(s) responsible :

Sponsor(s) :

Cooperation :

Publication(s) on the project : See above (research protocol).

Codes : A.10 - A.11 - B.6

THE GEORGE WASHINGTON UNIVERSITY MEDICAL CENTER. (2)

Original language : English

Title : **OCCUPATIONAL SAFETY AND HEALTH.**

Research protocol :

This project is one of social and scientific responsibility, with a need to monitor ongoing scientific data, place it in focus and, as scientists, make sure of the interpretation and usage of such information in the social and political arena. Articles such as in Ann. J. Industrial Med. 2: 273-291 (1981) and in Recent Advances in Cancer Control, Eds. Yamagata, Hirayama and Hisamichi, Excerpta Medica Internatl. Congress Series 622 (1983) review certain aspects of occupational safety and health problems.

Progress : start =   end =   publication =

Person(s) responsible :

Sponsor(s) :

Cooperation :

Publication(s) on the project : See above (research protocol).

Codes : A.10 - A.11 - B.6

122

INSTITUTE OF ENVIRONMENTAL HEALTH, UNIVERSITY OF CINCINNATI.

Original language : English

Title : **MAGNESIUM CONCENTRATION AS AN INDEX OF CHRYSOTILE EXPOSURE IN ASBESTOS REMOVAL.**

Research protocol :

Magnesium concentrations were determined by atomic absorption, asbestos fiber concentrations by National Institute for Occupational Safety & Health method 7400. Background concentrations were determined in University buildings. Two removal projects were sampled; the sample was split for count and magnesium analysis. Correlation between Mg concentration and fiber count was determined at each locations.

Progress : start = 06/06/85   end = 01/01/86   publication = 01/01/86

Person(s) responsible : Carl Mueller, Howard Ayer

Sponsor(s) : Internal departmental funds, NIOSH training grant.

Cooperation : National Institute for Occupational Safety & Health.

Publication(s) on the project :

M.S. Thesis, University of Cincinnati, Mr. Mueller.

Codes : A.12 - B.3 - C.1

JOHNS HOPKINS SCHOOL OF HYGIENE AND PUBLIC HEALTH

Original language : English

Title : **ASBESTOS AND LUNG CANCER CELL TYPE.**

Research protocol :

A nested case-control study was undertaken to investigate the relation-ship between asbestos exposure and lung cancer cell type. Cases were former employees of two Virginia shipyards, and were identified from the Virginia Tumor Registry. All cases were diagnosed with lung cancer between 1975-82. A stratified random sample of controls was selected from among former shipyard workers from the same two yards as the cases. The controls were selected from among former employees who resided in Virginia or died in the State between 1975-82. The sampling strata for selecting controls were defined by age, year, and shipyard of first employment and race. Two hundred ninety-eight cases, approximately equal proportions of squamous cell, small cell, large cell, and adenocarci-nomas, and four hundred twelve controls were traced for telephone interviews.

Job histories were abstracted from shipyard personnel records on all cases and controls and were the primary source of data used to derive measures of asbestos exposure. The questionnaire interview was used to obtain data on demographics, smoking history, shipyard employment history including reported asbestos exposure, asbestos exposure from work outside the shipyard, occupational exposure to known lung carcinogens, and history of selected diseases including lung cancer.

Analyses were conducted using the conditional maximum likelihood estimate of the odds ratio and logistic regression. The results from the analysis showed that adenocarcinoma had the strongest association with asbestos exposure and the only case group to be associated with a multiplicative interaction effect between asbestos exposure and smoking. The most significant associations were found for adenocarcinoma cases employed before 1950.

Progress : start =   end =   publication =

Person(s) responsible : Walter Stewart.

Sponsor(s) : Department of Energy.

Cooperation :

Publication(s) on the project :

Codes : A.10 - C.0

KUAKINI MEDICAL CENTER. (1)

Original language : English

Title : **ETHNICITY AND THE RESPONSE TO ASBESTOS**

Research protocol :

A multiethnic group of 231 active, former, and retired male asbestos workers was examined for the role of ethnicity in asbestos response based on radiological changes.

Although the frequency and distribution of radiological abnormalities were as expected in Caucasians, that of other ethnic group differed. Ethnicity modified the radiological pattern: in part by influencing asbestos dose, inorganic substances and smoking and perhaps in part by an independent action.

In conclusion, ethnicity may independently influence the development of asbestos induced disease.

Progress : start = 02/01/85   end = 12/01/85   publication =

Person(s) responsible : G. Fournier-Massey, D.G. Massey.

Sponsor(s) :

Cooperation : John A. Burns, School of Medicine of the University of Hawaii.

Publication(s) on the project :

Codes : A.9 - B.6 - C.0 - D.3

U.S.A.

KUAKINI MEDICAL CENTER. (2)

Original language : English

Title : **COST-EFFECTIVENESS IN ASBESTOS INVESTIGATION**.

Research protocol :

Radiological markers of asbestos-associated disease were found in 127 active shipyard workers on a screening of the 6022 member work force. Of these 117 were comprehensively examined for asbestos-associated disease and degree of pulmonary impairment.

The investigation was divided into 3 phases of increasing complexity: Phase 1 of consultation, PA, lateral, and 2 oblique radiographs, routine pulmonary function tests, sputum-cytology, and asbestos bodies, CBC, EKG. Phase 2 consisted of compliance and 2 levels of steady-slate exercise. Phase 3 included specialized investigation for cancer. Cost-effectiveness and cost benefit analysis was performed on the results.

In Phase 1, history, physical exam, 'B' reading of a PA radiographs and routine pulmonary function were cost effective. Induced sputum yielded higher results than home-collected for cytology. Phase 2, if indicated, should be performed in the same session as phase 1. One major advantage of the program was to allay the anxiety of the workers.

In conclusion, a comprehensive examination of those with asbestos markers can be less extensive than thought previously.

Progress : start = 1979   end = continuing   publication = to be submitted

Person(s) responsible : D.G. Massey, G. Fournier-Massey.

Sponsor(s) :

Cooperation : John A. Burns School of Medicine of the University of Hawaii.

Publication(s) on the project :

Codes : A.9 - B.6 - C.0 - D.3

MINE SAFETY AND HEALTH ADMINISTRATION (MSHA), U.S. DEPARTMENT OF LABOR.(1)

Original language : English

Title : **EVALUATION OF MICHAEL-WALTERS INDUSTRIES SEALANT, "STOPPIT".**

Research protocol :

"Stoppit", manufactured in the U.S. by Michael-Walters Industries, is a sodium silicate base sealant that contains asbestos. The sealant is used in the mining industry for building mine "stoppings" and is available in two grades: a trowel grade and a caulking grade. Limited studies with the trowel grade material showed that when dry, abrasion of the surface could cause for the release of asbestos fibers and the product breaks down when exposed to water.

Progress : start = 01/06/81   end = 21/09/81   publication =

Person(s) responsible : Thomas F. Tomb.

Sponsor(s) :

Cooperation :

Publication(s) on the project : Internal report within MSHA.

Codes : A.2 - A.12 - B.14 - C.0 - D.5.c) - E.37

U.S.A.

MINE SAFETY AND HEALTH ADMINISTRATION (MSHA), U.S. DEPARTMENT OF LABOR.(2)

Original language : English

Title : **INFORMATION ON THE ASBESTOS WASHER USED IN "MINE" SAFETY LAMPS.**

Research protocol :

The potential hazard associated with the use of an asbestos washer in mine safety lamps was briefly investigated. The washers contain Canadian Chrysotile Asbestos and pose a potential health problem when they are being cleaned during routine maintenance of the flame safety lamp. In limited studies the level of airborne asbestos was found to be well below the recommended control level of 2 fibers/cc greater than 5 micrometers in length. However, a recommendation was made that it would be prudent for the safety lamp manufacturers to use a substitute material.

Progress : start = 01/02/81   end = 22/05/81   publication =

Person(s) responsible : Thomas F. Tomb

Sponsor(s) :

Cooperation :

Publication(s) on the project : Letter to MSHA Headquarters.

Codes : A.2 - A.12 - B.12 - C.1 - D.11.a) - E.37

MOUNT SINAI SCHOOL OF MEDICINE. (1)

Original language : English

Title : **URINARY EXCRETION OF MODIFIED NUCLEOSIDES IN ASBESTOS INSULATION WORKERS.**

Research protocol :

During clinical field studies, 2,907 long-term asbestos insulation workers 1981-1983, urine sample were collected and then frozen. They are being analyzed for a variety of nucleosides. Preliminary studies has demonstrated significant modification of normal excretion patterns.

The mortality experience of the insulation workers is being ascertained, with 215 deaths in prospective observation 1981-1985. We will analyze the predictive significance of urinary nucleoside modifications in relation to subsequent risk of death of neoplasia.

Progress : start = 01/01/80   end = 31/12/87   publication =

Person(s) responsible : A. Fischbein.

Sponsor(s) : National Cancer Institute.

Cooperation : Colorado Cancer Center (E. Borek).

Publication(s) on the project :

Codes : A.10 - A.11 - D.3

MOUNT SINAI SCHOOL OF MEDICINE. (2)

Original language : English

Title : **STRESS AND RISK OF NEOPLASTIC ASBESTOS-ASSOCIATED DISEASE.**

Research protocol :

During clinical field examinations of 2,907 long-term asbestos insulation workers, extensive data were collected concerning psycho-social stress status. This is being updated yearly. The entire group of 2,907 men is being followed prospectively, and causes of deaths ascertained. By the fall of 1985, 215 deaths has occurred, a high proportion of asbestos-associated neoplasms. With additional experience, we will analyze expected and observed deaths by cause, in relation to baseline psychosocial stress status.

Progress : start = 01/08/80   end = 31/12/87   publication =

Person(s) responsible : L. Glickman.

Sponsor(s) : National Cancer Institute.

Cooperation : American Cancer Society.

Publication(s) on the project :

Codes : A.10 - A.11 - D.3

MOUNT SINAI SCHOOL OF MEDICINE. (3)

Original language : English

Title : **MANAGEMENT AND TREATMENT OF MESOTHELIOMA**

Research protocol :

> While a great deal of information has been obtained concerning epidemio-
> logical, clinical and pathological features of pleural and peritoneal
> mesothelioma, answers in managements and treatment have lagged behind. A
> special unit is being established at the Mount Sinai School of Medicine
> and the Mount Sinai Hospital to develop necessary research programs in
> this area.

Progress : start = 01/09/85   end =   publication =

Person(s) responsible : Alvin S. Teirstein, J.J. Selikoff.

Sponsor(s) : Mount Sinai School of Medicine.

Cooperation : Workers' Health Fund, AFL-CIO and others.

Publication(s) on the project :

Codes : A.9 - B.6

MOUNT SINAI SCHOOL OF MEDICINE. (4)

Original language : English

Title : **MORTALITY EXPERIENCE OF SHEET METAL WORKERS IN THE UNITED STATES AND CANADA.**

Research protocol :

> Asbestos-associated disease among sheet metal workers has been found in clinical cross-sectional studies. In addition, mesotheliomas have been observed in workers in this trade. Using a retrospective-prospective study design, we will investigate the mortality experience of sheet metal workers in the United States and Canada, members of the Sheet Metal Workers International Association.

Progress : start = 01/09/85   end = 01/12/86   publication =

Person(s) responsible : I.J. Selikoff, H. Seidman.

Sponsor(s) : Sheet Metal Workers International Association.

Cooperation : American Cancer Society.

Publication(s) on the project :

Codes : A.11 - D.3

MOUNT SINAI SCHOOL OF MEDICINE. (5)

Original language : English

Title : **ASBESTOS DISEASE AMONG SHEET METAL WORKERS IN THE UNITED STATES AND CANADA.**

Research protocol :

Problems of exposure during maintenance, repair, renovation, remodeling, particularly in the construction industry, have become increasingly more important. One group which seems to be particularly at risk are sheet metal workers. Clinical cross-sectional studies have demonstrated a significant prevalence of asbestosis among them. A systematic clinical field survey is being undertaken to determine the level of disease that may be found, to assist in the development of improved work practices, designed to eliminate or minimize exposure from asbestos materials in place.

Progress : start = 01/09/85   end = 31/12/86   publication =

Person(s) responsible : I.J. Selikoff.

Sponsor(s) : Sheet Metal Workers International Association.

Cooperation : American Cancer Society.

Publication(s) on the project :

Codes : A.11 - D.3

MOUNT SINAI SCHOOL OF MEDICINE. (6)

Original language : English

Title : **PREDICTIVE SIGNIFICANCE OF IMMUNOMODIFICATION IN ASBESTOS-ASSOCIATED DISEASE.**

Research protocol :

During 1981-1983, we examined 2,907 long-term asbestos insulation workers in the United States, in 19 cities. A good deal of information was obtained (clinical, radiological, physiological, biochemical). Extensive immunological studies were also done, including in vivo response to recall antigens and lymphocyte changes. These data were recorded. The 2,907 men involved are being followed prospectively and their mortality experience will be analyzed in terms of immunomodification findings.

Progress : start = 01/01/80   end = 31/12/87   publication =

Person(s) responsible : J.G. Bekesi, I.J. Selikoff, A. Fischbein.

Sponsor(s) : Mount Sinai School of Medicine.

Cooperation : American Cancer Society.

Publication(s) on the project :

Codes : A.9 - A.10 - B.6 - D.3

MOUNT SINAI SCHOOL OF MEDICINE. (7)

Original language : English

Title : **ASBESTOS DISEASE AMONG FAMILY CONTACTS OF AMOSITE ASBESTOS FACTORY WORKERS.**

Research protocol :

We are following prospectively all employees of the Union Asbestos and Rubber Company factory in Paterson, New Jersey, which operated 1941-1954. Their clinical and mortality experience is being recorded. Simultaneously, we established a registry of all family contacts who lived with these workers during their period of employment. Initial clinical and radiological findings were reported in 1979.

These family contacts, constituting a cohort, are being followed prospectively and their mortality experience is being ascertained.

Progress : start = 01/01/69   end = 31/12/86   publication =

Person(s) responsible : I.J. Selikoff, L. Joubert, H. Seidman.

Sponsor(s) : American Cancer Society.

Cooperation :

Publication(s) on the project :

Codes : A.10 - A.11 - C.2 - D.3

MOUNT SINAI SCHOOL OF MEDICINE. (8)

Original language : English

Title : **FAMILY CONTACTS OF ASBESTOS INSULATION WORKERS.**

Research protocol :

On January 1, 1943, there were 632 men who were members of Local 12 and Local 32 of the International Association of Health and Frost Insulators and Asbestos Workers in the New York - New Jersey Metropolitan area. The entire group has been followed since (approximately 70 are still alive at this writing). We ascertained which family members were living in their households 1942-1943. This group constitutes a cohort, which is being investigated in terms of their mortality experience since.

Progress : start = 01/01/65   end = 01/07/87   publication =

Person(s) responsible : I.J. Selikoff, L. Joubert, J.S. Kaffenburgh.

Sponsor(s) : American Cancer Society.

Cooperation :

Publication(s) on the project :

Codes : A.10 - D.3

MOUNT SINAI SCHOOL OF MEDICINE. (9)

Original language : English

Title : **CLINICAL FINDINGS AMONG LONG-TERM ASBESTOS INSULATION WORKERS IN THE UNITED STATES AND CANADA.**

Research protocol :

On January 1, 1967, there were 17,800 asbestos insulation workers, members of the International Association of Heat and Frost Insulators and Asbestos Workers, in the United States and Canada. On January 1, 1980, more than 4,000 workers were more than 30 years from onset of their work exposure, and were not known to have died. All were invited to come for clinical, radiological and physiological (and biochemical) examination. 2,907 attended survey clinics in Chicago, Columbus, Boston, Tampa, Atlanta, Baltimore, Syracuse, Albuquerque, Houston, Dallas, New Orleans, New Jersey. A great many findings were gathered and the entire group remains under observation as a separate cohort with the foregoing baseline data.

Progress : start = 01/01/80   end = 01/01/88   publication =

Person(s) responsible : I.J. Selikoff.

Sponsor(s) : American Cancer Society.

Cooperation : International Association of Heat and Frost Insulators and Asbestos Workers.

Publication(s) on the project :

Codes : A.10 - A.11 - D.3

MOUNT SINAI SCHOOL OF MEDICINE. (10)

Original language : English

Title : **PROGNOSTIC SIGNIFICANCE OF VITAMIN A AND β-CAROTENE BLOOD LEVELS FOR ASBESTOS-ASSOCIATED DISEASE.**

Research protocol :

During a clinical survey of long-term asbestos insulation workers in the United States, serum Vitamin A and β-carotene were measured. The entire group of 2,907 men is being followed prospectively for their mortality experience. Holding other variables constant, we are investigating the prognostic significance of variations in Vitamin A and β-carotene serum levels in relation to risk of death of asbestos-associated disease, including specific cancers.

Progress : start = 01/01/81   end = 31/12/87   publication =

Person(s) responsible : I.J. Selikoff

Sponsor(s) : American Cancer Society.

Cooperation : Hoffman-LaRoche.

Publication(s) on the project :

Codes : A.9 - A.10 - D.3

U.S.A.

MOUNT SINAI SCHOOL OF MEDICINE. (11)

Original language : English

Title : **DOSE-RESPONSE TO AMOSITE ASBESTOS EXPOSURE: PROSPECTIVE OBSERVATION.**

Research protocol :

932 men worked for shorter or longer periods at the Union Asbestos and Rubber Company in Paterson, New Jersey 1941-1954. Duration of employment is known and there is much information concerning levels of dust exposure. Serial observation is being maintained of the men who once worked in this factory, including radiological studies. Findings are being evaluated in terms of the dose (intensity x time) taking into account such variables as cigarette smoking, duration from onset, duration of exposure, work station, etc.

Progress : start = 01/01/54   end = 31/12/86   publication =

Person(s) responsible : I.J. Selikoff, M.D.

Sponsor(s) : American Cancer Society, National Institutes of Health.

Cooperation :

Publication(s) on the project :

Codes : A.10 - A.11 - C.2 - D.3

MOUNT SINAI SCHOOL OF MEDICINE. (12)

Original language : English

Title : **SHORT-TERM EXPOSURE AND LONG-TERM OBSERVATION OF AMOSITE ASBESTOS FACTORY WORKERS.**

Research protocol :

Because of wartime conditions, the duration of employment of the 932 men who worked for shorter or longer periods for the Union Asbestos and Rubber Company (amosite asbestos product) varied greatly. Some worked for as short as a day, others for one or more weeks, months and some, for 13 years until the plant closed in 1954. The entire cohort is being followed prospectively and the mortality experience is being evaluated in relation to the duration of employment (and associated calculated asbestos exposure).

Progress : start = 01/01/54   end = 31/12/86   publication =

Person(s) responsible : Herbert Seidman, I.J. Selikoff, M.D.

Sponsor(s) : American Cancer Society, National Institutes of Health.

Cooperation :

Publication(s) on the project : 1976 et seq.

Codes : A.10 - A.11 - C.2 - D.3

MOUNT SINAI SCHOOL OF MEDICINE. (13)

Original language : English

Title : **MORTALITY EXPERIENCE OF AMOSITE ASBESTOS FACTORY WORKERS; PROSPECTIVE OBSERVATION.**

Research protocol :

A factory (The Union Asbestos and Rubber Company) operated in Paterson, New Jersey, 1941-1954. They had manufactured a variety of insulation materials using amosite asbestos, from 1941 to 1945, and employed, for shorter or longer periods, 932 men. Altogether, 1,162 men worked until the plant closed in November, 1954. The entire cohort is being followed with causes of death ascertained as they occur. A large proportion of the men were examined during the 1950's and 1960's (and the survivors are still being followed clinically) so that smoking habits were ascertained.

Progress : start = 01/01/54   end =   publication =

Person(s) responsible : I.J. Selikoff, M.D.

Sponsor(s) : American Cancer Society; National Institutes of Health.

Cooperation :

Publication(s) on the project : 1961-1980 et seq.

Codes : A.10 - A.11 - C.2 - D.3

MOUNT SINAI SCHOOL OF MEDICINE. (14)

Original language : English

Title : **PROSPECTIVE CLINICAL OBSERVATION OF ASBESTOS INSULATION WORKERS.**

Research protocol :

On January 1, 1963, there were 1,249 asbestos insulation workers in the New York - Metropolitan area. 1,117 were examined (and the findings reported). The entire group of 1,249 has been followed prospectively with serial examinations and investigation of each death, as it occurred. Analyses are now underway to evaluate the prognostic significance of various clinical, physiological and roentgenological findings in 1963, taking into account demographic, smoking and other variables. In addition, we are investigating such things as rates of X-ray progression of asbestosis, pulmonary function changes, clinical disability.

Progress : start = 01/01/63   end =   publication = 01/07/86 et seq.

Person(s) responsible : I.J. Selikoff, M.D.

Sponsor(s) : American Cancer Society, National Institutes of Health.

Cooperation : Locals 12 and 32 of the International Association of Heat and Frost Insulators and Asbestos Workers.

Publication(s) on the project : 1964, 1965, and others.

Codes : A.10 - A.11 - D.3

MOUNT SINAI SCHOOL OF MEDICINE. (15)

Original language : English

Title : **MORTALITY EXPERIENCE OF ASBESTOS INSULATION WORKERS IN THE
UNITED STATES AND CANADA, PROSPECTIVE STUDY**

Research protocol :

17,800 members of the International Association of Heat and Frost
Insulators and Asbestos Workers, AFL-CIO, CLC were registered in a
prospective study on January 1, 1967. A good deal of information was
recorded for each. The entire cohort is being followed prospectively and
each death studied and the cause ascertained.

Progress : start = 01/01/67   end =   publication = 01/07/86 et seq.

Person(s) responsible : I.J. Selikoff, M.D.

Sponsor(s) : American Cancer Society, National Institutes of Health.

Cooperation : International Association of Heat and Frost Insulators and
Asbestos Workers.

Publication(s) on the project : Several

Codes : A.10 - A.11 - D.3

U.S.A.

NATIONAL INSTITUTE OF ENVIRONMENTAL HEALTH SCIENCES,
PULMONARY PATHOLOGY LABORATORY. (1)

Original language : English

Title : **PATHOGENESIS OF EARLY PULMONARY LESIONS INDUCED BY INHALED INORGANIC PARTICLES.**

Research protocol :

We have established animal models to elucidate the basic cellular mechanisms associated with alterations induced by inhalation of asbestos and silica. We have documented the initial deposition sites of the particles and the nature of the earliest lesions consequent to deposition. Autoradiography and ultrastructural morphometry demonstrate the cellular alterations which occur at alveolar duct bifurcations (i.e., sites of particle deposition) at varying times after a brief exposure (1 or 5 hrs) to chrysotile asbestos. The autoradiography showed that within 24 hrs after a 5-hr exposure, there was a highly significant increase in the number of terminal bronchiolar epithelial cells, proximal alveolar duct epithelial cells and alveolar duct epithelial cells which had incorporated tritiated thiymidine (3 HTdr) into nuclei. In addition, there were significant increases in the number of labelled interstitial cells in all three anatomic compartments. Significantly, morphometric studies of alveolar duct bifurcations showed that the volume as well as number of epithelial cells and macrophages were increased by 48 hrs after a 1-hr exposure to chrysotile asbestos. One month after this exposure, the numbers of interstital macrophages and fibroblasts were significantly increased. Moreover, the volume of the noncellular interstitial collagenous matrix was increased, signaling the presence of early asbestos-induced fibrogenesis. Further studies are ongoing to define the mechanisms through which the asbestos stimulates the fibrotic lung response.

Progress : start =   end = ongoing   publication =

Person(s) responsible : Arnold R. Brody, Ph. D.

Sponsor(s) : National Institute of Environmental Health Sciences.

Cooperation :

Publication(s) on the project :
1. Brody, A.R., Hill, L.H. Adkins, B., and O'Connor, R.W. Chrysotile asbestos inhalation in rats: deposition pattern and reaction of alveolar epithelium and pulmonary macrophages. Amer. Rev. Resp. Dis. 123:670-679, 1981.
2. Brody, A.R. and Hill, L.H. Interstitital accumulation of inhaled chrysotile asbestos fibers and consequent formation of microcalcifications. Amer. J. Pathol.: 109-107-114, 1982.
3. Brody, A.R., Roe, M.W., Evans J.N., and Davis, G.S. Deposition and translocation of inhaled silica in rats: quantification of macrophage participation and particle distribution in alveolar ducts. Lab. Invest. 47:533-42, 1982.

Codes : A.9.a) - A.9.b) - A.12 - B.7 - C.1

U.S.A.

NATIONAL INSTITUTE OF ENVIRONMENTAL HEALTH SCIENCES,
PULMONARY PATHOLOGY LABORATORY. (2)

Original language : English

Title : **ASBESTOS ACTIVATION OF COMPLEMENT-DEPENDENT CHEMOTACTIC FACTORS FOR MACROPHAGES.**

Research protocol :

Pulmonary macrophages migrate to the sites where inhaled chrysotile fibers initially are deposited (i.e., surfaces of alveolar duct bifurcations). These macrophages form a major component of an early asbestos-induced interstitial lesion in rats. To establish the basic cellular mechanisms of asbestos-induced lung disease, it is essential to determine the chemical mediators which attract macrophages to these sites of fiber disposition. Fluids, lavaged from the lungs of exposed rats contain substantial chemotactic activity for macrophages compared to fluids from sham-exposed animals. We hypothesize that this chemotactic activity is derived from complement activated by inhaled asbestos on alveolar surfaces. This contention is supported by observing that: Fractionation, by molecular sieve chromatography, of serum proteins and concentrated proteins lavaged from the lungs of asbestos-exposed rats showed that chemotactic activity was detected in the 14-18,000 MW range. This fractionation profile is similar to C5a, the chemotactic product of complement activation. In addition, rats treated with cobra venom factor (CVF) to deplete circulating complement as well as complement-deficient mice demonstrated significantly depressed macrophage accumulation. The complement-dependent chemotactic factor is activated during a 3-hr exposure to asbestos, but the chemotactic activity is not detectable by 8 days after exposure. Interestingly enough, when CVF-treated rats were exposed to asbestos, their macrophage response returned when circulating complement reached normal levels.

Progress : start =   end = ongoing   publication =

Person(s) responsible : Arnold R. Brody, Ph. D.

Sponsor(s) : National Institute of Environmental Health Sciences.

Cooperation : Pulmonary and Rheumatology Divisions, Duke University College of Medicine, Durham, NC.

Publication(s) on the project :
Warheit, D.B., Hill, L.H. and Brody, A.R.: Surface morphology and correlated phagocytic capacity of pulmonary macrophages lavaged from the lungs of rats. Exp. Lung Res., In Press, 1983.
Kouzan, S., Brody, A.R., et al. Production of arachidonic acid metabolites by pulmonary macrophages exposed in vitro to asbestos, carbonyl iron or calcium ionophore. Amer. Rev. Resp. Dis., 131: 624-632, 1985.
Warheit, D.B., George, G., Hill, L.H., Snyderman, R. and Brody, A.R., Inhaled asbestos activates a complement-dependent chemoattractant for macrophages. Lab. Invest., 52:505-514, 1985.

Codes : A.9.a) - A.9.b) - A.12 - B.7 - C.1

U.S.A.

NATIONAL INSTITUTE OF ENVIRONMENTAL HEALTH SCIENCES,
PULMONARY PATHOLOGY LABORATORY. (3)

Original language : English

Title : **INTERACTIONS OF INORGANIC PARTICLES WITH PULMONARY CELL MEMBRANES.**

Research protocol :

Non-specific (i.e., non-receptor mediated) binding and subsequent uptake
of positively-charged particles are mediated by negatively-charged cell
surface sialic acid groups. To support this hypothesis, we have shown
that chrysotile asbestos damages erythrocyte membranes through binding
to terminal sialic acid (SA) residues. The hemolytic events involved (1)
binding of the positively-charged chrysotile fibers to negatively-
charged SA groups, (2) rapid (within 5 min) distortion of the cells, (3)
redistribution of SA groups, and (4) alterations of intracellular Na^+,
K^+ ratios. Negatively-charged crocidolite asbestos bound to and
distorted red cells but had no effects on SA groups or ion flux. We have
extended such studies to pulmonary macrophages and have shown the
following: (1) Wheat germ agglutinin (WGA), a lectin which binds to
sialic acid, is distributed evenly across macrophage surface. (2)
Positively-charged carbonyl iron (Fe) spheres and chrysotile asbestos
fibers bind to macrophage membranes at 4°C, and the binding is blocked
by a dose-dependent pretreatment of the cells with WGA. Other lectins
such as Ricin and ConA do not inhibit binding at comparable doses. (3)
In the presence of WGA, over 90% of phagocytic activity is blocked, but
other lectins have no effect. These studies support our hypothesis that
charged surface sialic acid groups play a role in particle binding and
phagocytosis. In addition, ongoing studies have shown that particles
induce the production of arachidonic acid metabolites (potent mediators
of inflammation) through interactions between the terminal sialic acid
groups and particles bound to the macrophage membranes.

Progress : start =   end = ongoing   publication =

Person(s) responsible : Arnold R. Brody, Ph. D.

Sponsor(s) : National Institute of Environmental Health Sciences.

Cooperation :

Publication(s) on the project :
Brody, A.R. and Roe, M.W. Deposition pattern of inorganic particles at
the alveolar level in the lungs of rats and mice. Amer. Rev. Resp. Dis.,
128:724-729, 1983.
Brody, A.R., George, G., and Hill, L.H. Interactions of chrysotile and
crocidolite asbestos with red blood cell membranes: Chrysotile binds to
sialic acid. Lab. Invest., 49:468-475, 1983.
Kouzan, S., Gallagher, J., Elling, T. and Brody, A.R. Particle binding
to sialic acid residues on macrophage plasma membranes stimulates
arachidonic acid metabolism. Lab. Invest., 53:320-27, 1985.

Codes : A.9.a) - A.9.b) - A.12 - B.7 - C.1

NATIONAL INSTITUTE FOR OCCUPATIONAL SAFETY AND HEALTH (NIOSH).

Original language : English

Title : **ANALYTICAL METHODS FOR ASBESTOS FIBERS.**

Research protocol :

This project covers a number of areas related to asbestos aerosol sampling and analysis for industrial hygiene purposes. These areas include:
a. The investigation of electrostatic effects in sampling for asbestos aerosols through both a laboratory investigation of electrostatic losses during sampling and field measurements of charge of workplace aerosols.
b. Improvement of the NIOSH Method 7400 for asbestos fiber analysis.
c. The improvement of sample preparation for the NIOSH Proficiency Analytical Testing program, an interlaboratory quality assurance program.
d. The development of a Transmission Electron Microscope (TEM) method for determining asbestos aerosol concentration.
e. The completion of an evaluation of an automated technique for asbestos fiber counting with phase contrast light microscopy using the Magiscan image analyser.
f. The development of an automated fiber counting technique for TEM analysis.

Progress : start = 01/04/85   end = 30/09/89   publication =

Person(s) responsible : Paul A. Baron.

Sponsor(s) : NIOSH

Cooperation :

Publication(s) on the project :

Codes : A.7 − B.3 − B.6

WAYNE STATE UNIVERSITY, OCCUPATIONAL AND ENVIRONMENTAL HEALTH.

Original language : English

Title : **WATER POLLUTION STUDIES FOR RESERVE MINING COMPANY.**

Research protocol :

Study on Municipal Waters in Selected Communities

Tap water samples were collected in various municipalities on and off Lake Superior, and the particulate contaminants were studied by x-ray diffraction analysis and electron microscopy including microprobe. Qualitative and quantitative comparisons of these municipal waters, specially with respect to fibrous particles, were obtained.

Study of the Origins of the Particulate Contaminants in Duluth Tap Water

The west end of Lake Superior receives numerous tributaries of natural, municipal, or industrial origin which contribute to the pollution and/or eutrophication of the lake in various proportions. Water samples were collected at the mouths of these tributaries and at various locations in the lake in order to determine the extent to which each of these contributes to the contamination of the water at the municipal water intake at Duluth.

Progress : start = 17/07/73   end = 31/12/73   publication =

Person(s) responsible : Andrew L. Reeves, Neil D. Krivanek.

Sponsor(s) : Reserve Mining Company.

Cooperation : Illinois Institute of Technology.

Publication(s) on the project : Report to the sponsor.

Codes : A.3 - A.5 - A.12 - B.3 - B.10 - C.2 - C.6.

USSR

INSTITUTE OF GEOLOGY AND GEOPHYSICS,
ACADEMY OF SCIENCES OF THE USSR, SIBERIAN BRANCH.

Original language : Russian

Title : **THE REFINEMENT OF THE SPACE GROUP AND PHASE TRANSITION
IN CHRYSOTILE-ASBESTOS**

Research protocol :

The chrysotile-asbestos structure according to powder diffraction data refers to monoclinic classes (PDF-21-543) and is characterized by three space groups with the same rules of systematic absences: C^3_s - Cm; C^3_2 - C2; C^3_{2h} - C2/M.

Phase transition is set at T - 40°C. Vibrational spectra analyses using a previously suggested method (Arkhipenko, Bokiy, Kristal (1977), 22.G.1176) showed the following:
1. The space group between the groups with the same rules of systematic absences for the initial structure is defined as centrosymmetrical - C^3_{2h} - C2/m.
2. Inversion centre loss by phase transition is deduced (space group: C^3_s - CM or C^3_2 - C2).
3. The increasing brittling by thermal drop to T - 40°C with the change of its structure due to phase transition is explained.

Progress : start =   end =   publication =

Person(s) responsible : Arkhipenko D.K., Bokyi G.B., Palchik N.A.,
                        Simonov V.A.

Sponsor(s) :

Cooperation :

Publication(s) on the project : Mineralogical Journal (Kiev USSR)

Codes : A.4 - A.5 - C.5.a)

Section 2

CLASSIFICATION OF STUDIES BY SUBJECT

MULTI-DISCIPLINARY   A.0

National Occupational Health & Safety Commission	Australia
Österreichische Staub (Silikose)-Bekämpfungsstelle	Austria
Micro & Trace Analysis Centre (MITAC), University of Antwerp (U.I.A.)	Belgium
Service de Pneumologie, Hôpital Erasme	Belgium
Centre spécialisé en technologie minérale	Canada
Department of Soil Science, University of British Columbia	Canada
Laboratoire de caractérisation de l'amiante (LCA), Faculté des Sciences, Université de Sherbrooke	Canada
Centre de Recherches sur la Physico-Chimie des Surfaces Solides (CNRS) (1)	France
Ecole Supérieure de l'Energie et des Matériaux, Université d'Orléans	France
Istituto Superiore di Sanità - Laboratorio di Ultrastrutture	Italy
Academy of Mining and Metallurgy, Interbranch Institute of Building and Refractory Materials	Poland
Construction Materials Research Group, Civil Engineering Department, University of Surrey	United Kingdom
Former Aerosol Laboratory, University of Essex	United Kingdom
Water Research Centre Engineering	United Kingdom

GEOLOGY   A.1

LA 10 - CNRS & IOPG	France
Institut für Mineralogie, Ruhr-Universität Bochum	Germany (FRG)
Anjalena Publications Ltd. (1)	United Kingdom

MINING   A.2

Centre spécialisé en technologie minérale   Canada

Institute of Tuberculosis and Respiratory Diseases   Czechoslovakia

Department of Mineral and Energy Affairs,   South Africa
Air Quality Research

Anjalena Publications Ltd. (1)   United Kingdom

Mine Safety and Health Administration (MSHA),   U.S.A.
U.S. Department of Labor (1) (2)

CHEMISTRY   A.3

Micro & Trace Analysis Centre (MITAC), University   Belgium
of Antwerp (U.I.A.)

Centre spécialisé en technologie minérale   Canada

Département de Chimie, Université Laval   Canada

Département de Chimie, Université de Sherbrooke   Canada

Department of Soil Science, University of British   Canada
Columbia

Laboratoire de caractérisation de l'amiante,   Canada
Faculté des Sciences, Université de Sherbrooke

Laboratory of Catalysis, Department of Chemistry,   Canada
Concordia University,

Programme de recherche sur l'amiante, Université de   Canada
Sherbrooke (1)(2)(3)

Centre de Recherches sur la Physico-Chimie des   France
Surfaces Solides (CNRS) (2)

LA 10 - CNRS & IOPG   France

Institut für Mineralogie, Ruhr-Universität Bochum   Germany (FRG)

Laboratory of Biochemistry and Laboratory of   Israel
Cellular Immunology, Carmel Hospital

Anjalena Publications Ltd. (1)   United Kingdom

Wayne State University, Occupational and Environmental   U.S.A
Health

PHYSICS   A.4

Centre spécialisé en technologie minérale   Canada

Department of Physics, Dalhousie University   Canada

Laboratoire de caractérisation de l'amiante (LCA),   Canada
Faculté des Sciences, Université de Sherbrooke

LA 10 - CNRS & IOPG   France

Institut für Mineralogie, Ruhr-Universität Bochum   Germany (FRG)

Anjalena Publications Ltd. (1)   United Kingdom

Health & Safety Executive Research and Laboratory   United Kingdom
Services Division

Institute of Geology and Geophysics, Academy of   U.S.S.R.
Sciences of the USSR, Siberian Branch

MINERALOGY   A.5

Micro & Trace Analysis Centre (MITAC), University   Belgium
of Antwerp (U.I.A.)

Service de Pneumologie, Hôpital Erasme   Belgium

Centre spécialisé en technologie minérale   Canada

Programme de recherche sur l'amiante,   Canada
Université de Sherbrooke (6)

Institut National de Recherche sur la Sécurité   France
(I.N.R.S.) (2)

LA 10 - CNRS & IOPG   France

Institut für Mineralogie, Ruhr-Universität Bochum   Germany (FRG)

Centro di Studio per i Problemi Minerari   Italy
c/o Dipartimento Georisorse e Territorio - Politecnico

Centro Studi e Ricerche sugli Effetti biologici delle   Italy
Polveri inalate - Istituto di Medicina del Lavoro -
Università di Milano (4)

Academy of Mining and Metallurgy, Interbranch   Poland
Institute of Building and Refractory Materials

Service Cantonal d'Ecotoxicologie / Genève   Switzerland

Anjalena Publications Ltd. (1)   United Kingdom

Wayne State University, Occupational and   U.S.A.
Environmental Health

Institute of Geology and Geophysics, Academy of   U.S.S.R.
Sciences of the USSR, Siberian Branch

"MINERALURGY"   A.6

Département de Chimie, Université de Sherbrooke   Canada

"MINERALURGY" - recovery   A.6.a)

Woodsreef Mines Limited (2)   Australia

Cassiar Mining Corporation   Canada

"MINERALURGY" - classification   A.6.b)

Cassiar Mining Corporation   Canada

Programme de recherche sur l'amiante,   Canada
Université de Sherbrooke (3)

METROLOGY A.7

Service de Pneumologie, Hôpital Erasme	Belgium
Département de Chimie, Université Laval	Canada
Institut National de Recherche sur la Sécurité (I.N.R.S.) (1)	France
I.N.S.E.R.M., Unité 139 (5)	France
Laboratoire d'Etude des Particules Inhalées (LEPI) (1) (3)	France
Berufsgenossenschaftliches Institut für Arbeitssicherheit - BIA	Germany (FRG)
Institut Universitaire de Médecine du Travail et d'Hygiène Industrielle	Switzerland
Health & Safety Executive Research and Laboratory Services Division	United Kingdom
National Institute for Occupational Safety and Health (NIOSH)	U.S.A.

MECHANICS A.8

Sydney University, Department of Mechanical Engineering	Australia
Construction Materials Research Group, Civil Engineering Department, University of Surrey	United Kingdom

BIOLOGY

Centre spécialisé en technologie minérale	Canada
Department of Soil Science, University of British Columbia	Canada
Programme de recherche sur l'amiante, Université de Sherbrooke (1)	Canada
Pneumoconiosis Research Unit, School of Public Health, West China University of Medical Sciences (2)	China
Institute of Occupational Health (2)	Finland
Clinique toxicologique, Hôpital Fernand-Widal	France
Unité de Recherches Biomathématiques et Biostatistiques, INSERM U 263, Université Paris 7	France
Laboratory of Biochemistry and Laboratory of Cellular Immunology, Carmel Hospital	Israel
Toyama Medical and Pharmaceutical University, Department of Community Medicine	Japan
Kuakini Medical Center (1)(2)	U.S.A.
Mount Sinai School of Medicine (3) (6) (10)	U.S.A.

BIOLOGY - experimental

Faculté de médecine/Département de pneumologie, Université de Sherbrooke	Canada
Laboratoire de Biochimie et de Toxicologie Pulmonaires, Département de Biologie, Faculté des Sciences, Université de Sherbrooke	Canada
Laboratoire de caractérisation de l'amiante, (LCA) Faculté des Sciences, Université de Sherbrooke	Canada
Institute of Occupational Health (2)	Finland
I.N.S.E.R.M. Unité 139 (1) (2) (3) (4) (5) (6)	France
Department of Community Medicine, Toyama Medical and Pharmaceutical University	Japan

Department of Hygiene, Kawasaki Medical School	Japan
Institute of Occupational Medicine (1)(2)(3)(4)(5)(6)	United Kingdom
Strangeways Research Laboratory	United Kingdom
National Institute of Environmental Health Sciences, Pulmonary Pathology Laboratory (1)(2)(3)	U.S.A.

BIOLOGY - bio-medicine   A.9.b)

Department of Physics, Dalhousie University	Canada
Faculté de médecine/Département de pneumologie, Université de Sherbrooke	Canada
Pneumoconiosis Research Unit, School of Public Health, West China University of Medical Sciences (1)	China
Institute of Occupational Health (1)(2)(3)(4)	Finland
Centro Studi e Ricerche sugli Effetti biologici delle Polveri inalate - Istituto di Medicina del Lavoro - Università di Milano (1)(2)(3)	Italy
Istituto Superiore di Sanità - Laboratorio di Ultrastrutture	Italy
Department of Community Medicine, Toyama Medical and Pharmaceutical University	Japan
National Institute of Environmental Health Sciences, Pulmonary Pathology Laboratory (1)(2)(3)	U.S.A.

Asbestos Research Group, Queen Elizabeth II Medical Centre (1)(2)	Australia
Institute of Environmental Hygiene, University of Vienna (1)(2)	Austria
Micro & Trace Analysis Centre (MITAC), University of Antwerp (U.I.A.)	Belgium
Service de Pneumologie, Hôpital Erasme	Belgium
Département de Chimie, Université de Sherbrooke	Canada
Institut Armand-Frappier/Centre de recherche en épidémiologie (1)(2)	Canada
Ontario Ministry of Labour	Canada
School of Occupational Health, McGill University	Canada
Research Unit of Pneumoconiosis, School of Public Health, West China Medical University (2)	China
Clinique toxicologique, Hôpital Fernand-Vidal	France
Institut National de Recherche sur la Sécurité (I.N.R.S.) (2)	France
I.N.S.E.R.M. Unité 139 (6)(7)(8)(9)	France
Laboratoire d'Etude des Particules Inhalées (LEPI) (2)	France
Unité de Recherches Biomathématiques et Biostatistiques, INSERM U 263, Université Paris 7	France
Pulmologische Abteilung, Krankenhaus Wittenberg-Apollensdorf	Germany (GDR)
Centro Studi e Ricerche sugli Effetti biologici delle Polveri inalate - Istituto di Medicina del Lavoro - Università di Milano (4)	Italy
Istituto Superiore di Sanità - Laboratorio di Ultrastrutture	Italy
Servizio di Anatomia e Istologia Patologica, Ospedale di Monfalcone	Italy
Department of Community Medicine, Toyama Medical and Pharmaceutical University	Japan
Department of Oncology, University Hospital Umeå	Sweden

Clinic of Occupational Medicine	Sweden
Health & Safety Executive, Epidemiology and Medical Statistics Unit	United Kingdom
Institute of Occupational Medicine (7)	United Kingdom
MRC Environmental Epidemiology Unit	United Kingdom
Respiratory Investigation Centre, Belfast City Hospital	United Kingdom
The George Washington University Medical Center (1)	U.S.A.
Johns Hopkins School of Hygiene and Public Health	U.S.A.
Mount Sinai School of Medicine (1)(2)(6)(7)(8)(9) (10)(11)(12)(13)(14)(15)	U.S.A

INDUSTRIAL MEDICINE   A.11

Institute of Environmental Hygiene, University of Vienna (1)	Austria
Département de Chimie, Université de Sherbrooke	Canada
Ontario Ministry of Labour	Canada
Programme de recheche sur l'amiante, Université de Sherbrooke (2)	Canada
School of Occupational Health, McGill University	Canada
Institute of Tuberculosis and Respiratory Diseases	Czechoslovakia
Institute of Occupational Health (1)(2)(3)(4)	Finland
Institut National de Recherche sur la Sécurité (I.N.R.S.) (2)	France
Beral Bremsbelag GmbH	Germany (FRG)
Pulmologische Abteilung, Krankenhaus Wittenberg-Apollensdorf	Germany (GDR)
Environmental Pollution Research Centre	India

Hebrew University, Department of Medical Ecology	Israel
Laboratory of Biochemistry and Laboratory of Cellular Immunology, Carmel Hospital	Israel
Centro Studi e Ricerche sugli Effetti biologici delle Polveri inalate - Istituto di Medicina del Lavoro - Università di Milano (1)(2)(3)	Italy
Istituto Superiore di Sanità - Laboratorio Ultrastrutture	Italy
Italian Railway Health Service	Italy
Department of Hygiene, Kawasaki Medical School	Japan
Clinic of Occupational Medicine	Sweden
University of Uppsala, Department of Lung Medicine	Sweden
Health & Safety Executive, Epidemiology and Medical Statistics Unit	United Kingdom
Institute of Occupational Medicine (8)	United Kingdom
MRC Environmental Epidemiology Unit	United Kingdom
Pathology Department, Manchester University	United Kingdom
The George Washington University Medical Center (1)	U.S.A.
Mount Sinai School of Medicine (1)(2)(4)(5)(7)(9)(11)(12)(13)(14)(15)	U.S.A.

ENVIRONMENT   A.12

Woodsreef Mines Limited (1)(2)   Australia

Institute of Environmental Hygiene,   Austria
University of Vienna (2)

Micro & Trace Analysis Centre (MITAC),   Belgium
University of Antwerp (U.I.A.)

Centre spécialisé en technologie minérale   Canada

Département de Chimie, Université de Sherbrooke   Canada

Department of Community Health Sciences,   Canada
Faculty of Medicine, The University of Calgary

Department of Soil Science, University of British   Canada
Columbia

Laboratoire de Biochimie et de Toxicologie   Canada
Pulmonaires, Département de Biologie,
Faculté des Sciences, Université de Sherbrooke

Laboratoire de caractérisation de l'amiante (LCA),   Canada
Faculté des Sciences, Université de Sherbrooke

Ontario Research Foundation   Canada

Programme de recherche sur l'amiante,   Canada
Université de Sherbrooke (5)

Research Unit of Pneumoconiosis, School of Public   China
Health, West China Medical University (2)

Centro di Studio per i Problemi Minerari   Italy
c/o Dipartimento Georisorse e Territorio-Politecnico

Centro Studi e Ricerche sugli Effetti biologici delle   Italy
Polveri inalate - Istituto di Medicina del Lavoro -
Università di Milano (1)(4)

Istituto Superiore di Sanità - Laboratorio di   Italy
Ultrastrutture

Italian Railway Health Service   Italy

Asahi Glass Co., Ltd., Research Laboratory   Japan

Department of Hygiene, Kawasaki Medical School   Japan

Department of Mineral and Energy Affairs,   South Africa
Air Quality Research

National Board of Occupational Safety and   Sweden
Health (1)(2)(3)

Institut Universitaire de Médecine du Travail et d'Hygiène Industrielle	Switzerland
Anjalena Publications Ltd. (1)	United Kingdom
Health & Safety Executive Research Laboratory Services Division	United Kingdom
East Bay Municipal Utility District,	U.S.A.
Institute of Environmental Health, University of Cincinnati	U.S.A.
Mine Safety and Health Administration (MSHA), U.S. Department of Labor (1)(2)	U.S.A.
National Institute of Environmental Health Sciences, Pulmonary Pathology Laboratory (1)(2)(3)	U.S.A.
Wayne State University, Occupational and Environmental Health	U.S.A

ECONOMY   A.13

Département de Chimie, Université de Sherbrooke	Canada
LA 10 - CNRS & IOPG	France

LEGISLATION   A.14

Asahi Glass Co., Ltd., Research Laboratory	Japan
Institute of Occupational Medicine (8)	United Kingdom

OTHER DISCIPLINES   A.16

Programme de recherche sur l'amiante, Université de Sherbrooke (4)	Canada
Pulmologische Abteilung, Krankenhaus Wittenberg-Apollensdorf (medical history)	Germany (GDR)
The George Washington University Medical Center (2) (social and political aspects)	U.S.A.

MULTI-TECHNIQUES   B.0

Institute of Environmental Hygiene, University of Vienna (2)	Austria
Micro & Trace Analysis Centre (MITAC), University of Antwerp (U.I.A.)	Belgium
Department of Soil Science, University of British Columbia	Canada
Programme de recherche sur l'amiante, Université de Sherbrooke (1)(3)	Canada
Laboratoire d'Etude des Particules Inhalées (LEPI) (1)(3)	France
Institut für Mineralogie, Ruhr-Universität Bochum	Germany (FRG)
Istituto Superiore di Sanità - Laboratorio di Ultrastrutture	Italy
Department of Hygiene, Kawasaki Medical School	Japan
Anjalena Publications Ltd. (1)	United Kingdom
Construction Materials Research Group, Civil Engineering Department, University of Surrey	United Kingdom

PROSPECTING   B.1

LA 10 - CNRS & IOPG	France

EXTRACTION   B.2

Cassiar Mining Corporation	Canada
Department of Soil Science, University of British Columbia	Canada
Pathology Department, Manchester University	United Kingdom

SAMPLING AND CLASSIFICATION

Österreichische Staub (Silikose)-Bekämpfungsstelle	Austria
Département de Chimie, Université Laval	Canada
Department of Soil Science, University of British Columbia	Canada
Ontario Research Foundation	Canada
Institut National de Recherche sur la Sécurité (I.N.R.S.) (1)(2)	France
Centre de Recherches sur la Physico-Chimie des Surfaces Solides (CNRS) (1)	France
Istituto Superiore di Sanità - Laboratorio di Ultrastrutture	Italy
National Board of Occupational Safety and Health (1)(2)(3)	Sweden
Former Aerosol Laboratory, University of Essex	United Kingdom
Health & Safety Executive Research and Laboratory Services Division	United Kingdom
Pathology Department, Manchester University	United Kingdom
Institute of Environmental Health, University of Cincinnati	U.S.A
National Institute for Occupational Safety and Health (NIOSH)	U.S.A
Wayne State University, Occupational and Environmental Health	U.S.A.

PROCESSING   B.4

Woodsreef Mines Limited (1)(2)   Australia

Cassiar Mining Corporation   Canada

Institute of Tuberculosis and Respiratory Diseases   Czechoslovakia

Centre de Recherches sur la Physico-Chimie des   France
Surfaces Solides (CNRS) (2)

LA 10 - CNRS & IOPG   France

Italian Railway Health Service   Italy

Asahi Glass Co., Ltd., Research Laboratory   Japan

DUST CONTROL   B.5

Woodsreef Mines Limited (1)(2)   Australia

Österreichische Staub (Silikose)-Bekämpfungsstelle   Austria

Cassiar Mining Corporation   Canada

Centre spécialisé en technologie minérale   Canada

Département de Chimie, Université Laval   Canada

Programme de recherche sur l'amiante,   Canada
Université de Sherbrooke (5)

Research Unit of Pneumoconiosis, School of Public   China
Health, West China Medical University (2)

Institut National de Recherche sur la Sécurité   France
(I.N.R.S.) (1)(2)

Laboratoire d'Etude des Particules Inhalées   France
(LEPI)(2)(3)

Istituto Superiore di Sanità - Laboratorio di   Italy
Ultrastrutture

Italian Railway Health Service   Italy

National Occupational Health & Safety Commission	Australia
Institute of Environmental Hygiene, University of Vienna (1)(2)	Austria
Service de Pneumologie, Hôpital Erasme	Belgium
Research Unit of Pneumoconiosis, School of Public Health, West China Medical University (2)	China
Institute of Tuberculosis and Respiratory Diseases	Czechoslovakia
Institute of Occupational Health (1)(2)(3)(4)	Finland
Laboratoire d'Etude des Particules Inhalées (LEPI)(2)	France
Environmental Pollution Research Centre	India
Hebrew University, Department of Medical Ecology	Israel
Laboratory of Biochemistry and Laboratory of Cellular Immunology, Carmel Hospital	Israel
Istituto Superiore di Sanità - Laboratorio di Ultrastrutture	Italy
Servizio di Anatomia e Istologia Patologica, Ospedale di Monfalcone	Italy
University of Uppsala, Department of Lung Medicine	Sweden
Institute of Occupational Medicine (8)	United Kingdom
Pathology Department, Manchester University	United Kingdom
The George Washington University Medical Center (1)	U.S.A.
Kuakini Medical Center (1)(2)	U.S.A.
Mount Sinai School of Medicine (3)(6)	U.S.A.
National Institute for Occupational Safety and Health (NIOSH)	U.S.A.

ANIMAL EXPERIMENTS   B.7

Research Unit of Pneumoconiosis, School of Public   China
Health, West China Medical University (1)(2)

Institute of Occupational Health (2)   Finland

Institute of Occupational Medicine (1)(2)(3)(4)(5)(6)   United Kingdom

Strangeways Research Laboratory   United Kingdom

National Institute of Environmental Health Sciences,   U.S.A.
Pulmonary Pathology Laboratory (1)(2)(3)

HANDLING   B.8

Cassiar Mining Corporation   Canada

TRANSPORT   B.8.a)

Cassiar Mining Corporation   Canada

PACKAGING/LABELLING   B.8.b)

Woodsreef Mines Limited (2)   Australia

Cassiar Mining Corporation   Canada

QUALITY CONTROL   B.9

Cassiar Mining Corporation   Canada

Programme de recherche sur l'amiante,   Canada
Université de Sherbrooke (3)(4)(6)

Istituto Superiore di Sanità - Laboratorio di   Italy
Ultrastrutture

Service Cantonal d'Ecotoxicologie/Genève   Switzerland

WASTE MANAGEMENT   B.10

Woodsreef Mines Limited (2)   Australia

Centre spécialisé en technologie minérale   Canada

Department of Soil Science, University of British   Canada
Columbia

Istituto Superiore di Sanità - Laboratorio di   Italy
Ultrastrutture

East Bay Municipal Utility District   U.S.A.

Wayne State University, Occupational and   U.S.A
Environmental Health

WASTE DISPOSAL   B.10.a)

Cassiar Mining Corporation   Canada

RECYCLING   B.11

Centre spécialisé en technologie minérale   Canada

Mine Safety and Health Administration (MSHA),   U.S.A.
U.S. Department of Labor (2)

ENERGY CONSUMPTION   B.12

Cassiar Mining Corporation   Canada

Department of Community Health Sciences, Faculty   Canada
of Medicine, The University of Calgary

LA 10 - CNRS & IOPG   France

MARKETING   B.13

Woodsreef Mines Limited (1)   Australia

Cassiar Mining Corporation   Canada

OTHER TECHNIQUES   B.14

Laboratory of Catalysis, Department of Chemistry,   Canada
Concordia University

Department of Oncology, University Hospital Umeå   Sweden

Institute of Occupational Medicine (7)   United Kingdom

Mine Safety and Health Administration (MSHA),   U.S.A.
U.S. Department of Labor (1)

ANY FIBRES   C.0

Österreichische Staub (Silikose)-Bekämpfungsstelle	Austria
Micro & Trace Analysis Centre (MITAC), University of Antwerp (U.I.A.)	Belgium
Department of Physics, Dalhousie University	Canada
Faculté des Sciences/Département de Biochimie, Université Laval	Canada
Institut Armand-Frappier/Centre de Recherche en Epidémiologie (1)(2)	Canada
Laboratoire de caractérisation de l'amiante, Faculté des Sciences, Université de Sherbrooke	Canada
Programme de recherche sur l'amiante, Université de Sherbrooke (2)(4)	Canada
Berufsgenossenschaftliches Institut für Arbeitssicherheit - BIA	Germany (FRG)
Hebrew University, Department of Medical Ecology	Israel
Centro di Studio per i Problemi Minerari c/o Dipartimento Georisorse e Territorio-Politecnico	Italy
Istituto Superiore di Sanità - Laboratorio di Ultrastrutture	Italy
Servizio di Anatomia e Istologia Patologica, Ospedale di Monfalcone	Italy
Department of Hygiene, Kawasaki Medical School	Japan
National Board of Occupational Safety and Health (1)(2)(3)	Sweden
Health & Safety Executive, Epidemiology and Medical Statistics Unit	United Kingdom
Health & Safety Executive, Research and Laboratory Services Division	United Kingdom
Institute of Occupational Medicine (7)(8)	United Kingdom
Water Research Centre Engineering	United Kingdom
Johns Hopkins School of Hygiene and Public Health	U.S.A.
Kuakini Medical Center (1)(2)	U.S.A.
Mine Safety and Health Administration (MSHA), U.S. Department of Labor (1)	U.S.A.

Woodsreef Mines Limited (1)(2)	Australia
Institute of Environmental Hygiene, University of Vienna (1)	Austria
Département de Chimie, Université Laval	Canada
Department of Community Health Sciences, Faculty of Medicine, The University of Calgary	Canada
Department of Soil Science, University of British Columbia	Canada
Faculté de médecine/Départment de pneumologie, Université de Sherbrooke	Canada
Laboratoire de Biochimie et de Toxicologie Pulmonaires, Département de Biologie, Faculté des Sciences, Université de Sherbrooke	Canada
Laboratoire de caractérisation de l'amiante, Faculté des Sciences, Université de Sherbrooke	Canada
Laboratory of Catalysis, Department of Chemistry, Concordia University	Canada
Ontario Ministry of Labour	Canada
Programme de recherche sur l'amiante, Université de Sherbrooke (1)(3)(5)	Canada
School of Occupational Health, McGill University	Canada
Pneumoconiosis Research Unit, School of Public Health, West China University of Medical Sciences (1)	China
Institute of Tuberculosis and Respiratory Diseases	Czechoslovakia
Institute of Occupational Health (2)	Finland
Ecole Supérieure de l'Energie et des Matériaux, Université d'Orléans	France
I.N.S.E.R.M. Unité 139 (1)(2)(3)	France
Centre de Recherches sur la Physico-Chimie des Surfaces Solides (CNRS) (1)(2)	France
Laboratoire d'Etude des Particules Inhalées (LEPI)(2)(3)	France
Academy of Mining and Metallurgy, Interbranch Institute of Building and Refractory Materials	Poland
Service Cantonal d'Ecotoxicologie/Genève	Switzerland

Anjalena Publications Ltd. (1)   United Kingdom

Former Aerosol Laboratory, University of Essex   United Kingdom

Institute of Occupational Medicine (1)(2)(3)(5)(6)   United Kingdom

MRC Environmental Epidemiology Unit   United Kingdom

Strangeways Research Laboratory   United Kingdom

Institute of Environmental Health, University of   U.S.A.
Cincinnati

Mine Safety and Health Administration (MSHA),   U.S.A
U.S. Department of Labor (2)

National Institute of Environmental Health   U.S.A.
Sciences, Pulmonary Pathology Laboratory (1)(2)(3)

AMPHIBOLE   C.2

Institute of Environmental Hygiene, University of   Austria
Vienna (1)(2)

Laboratoire de Biochimie et de Toxicologie   Canada
Pulmonaires, Département de Biologie,
Faculté des Sciences, Université de Sherbrooke

Ontario Ministry of Labour   Canada

Pneumoconiosis Research Unit, School of Public Health,   China
West China University of Medical Sciences (1)(2)

Institute of Occupational Health (2)   Finland

Laboratoire d'Etude des Particules Inhalées (LEPI)(2)(3)   France

Centre de Recherches sur la Physico-Chimie des   France
Surfaces Solides (CNRS)(1)

Institut für Mineralogie, Ruhr-Universität Bochum   Germany (FRG)

Academy of Mining and Metallurgy, Interbranch   Poland
Institute of Building and Refractory Materials

Anjalena Publications Ltd.   United Kingdom

Former Aerosol Laboratory, University of Essex   United Kingdom

Institute of Occupational Medicine (3)(6)   United Kingdom

Mount Sinai School of Medicine (7)(11)(12)(13)   U.S.A.
Wayne State University, Occupational and   U.S.A.
Environmental Health

MODIFIED FIBRES   <u>C.3</u>

Woodsreef Mines Limited (1)(2)   Australia

Micro & Trace Analysis Centre (MITAC),   Belgium
University of Antwerp (U.I.A.)

Laboratoire de caractérisation de l'amiante (LCA),   Canada
Faculté des Sciences, Université de Sherbrooke

Programme de recherche sur l'amiante, Université   Canada
de Sherbrooke (1)(3)(6)

I.N.S.E.R.M. Unité 139 (1)(3)(4)   France

ARTIFICIAL FIBRES   <u>C.4</u>

Micro & Trace Analysis Centre (MITAC),   Belgium
University of Antwerp (U.I.A.)

Faculté de médecine/Département de pneumologie,   Canada
Université de Sherbrooke

Laboratoire de Biochimie et de Toxicologie   Canada
Pulmonaires, Département de Biologie,
Faculté des Sciences, Université de Sherbrooke

Laboratoire de caractérisation de l'amiante (LCA),   Canada
Faculté des Sciences, Université de Sherbrooke

Institut National de Recherche sur la Sécurité   France
(I.N.R.S.) (1)(2)

Institut für Mineralogie, Ruhr-Universität Bochum   Germany (FRG)

Asahi Glass Co., Ltd., Research Laboratory   Japan

OTHER FIBRES   C.5

National Occupational Health & Safety Commission   Australia

Sydney University, Department of Mechanical Engineering Australia

Micro & Trace Analysis Centre (MITAC),   Belgium
University of Antwerp (U.I.A.)

Faculté de médecine/Département de pneumologie,   Canada
Université de Sherbrooke

OTHER NATURAL FIBRES   C.5.a)

Asbestos Research Group,   Australia
Queen Elizabeth II Medical Centre (1)(2)

Laboratoire de Biochimie et de Toxicologie   Canada
Pulmonaires, Département de Biologie,
Faculté des Sciences, Université de Sherbrooke

Laboratoire de caractérisation de l'amiante (LCA),   Canada
Faculté des Sciences, Université de Sherbrooke

Institut National de Recherche sur la Sécurité   France
(I.N.R.S.)(1)

I.N.S.E.R.M. Unité 139 (1)(3)   France

Institute of Occupational Medicine (4)   United Kingdom

Institute of Geology and Geophysics, Academy of   U.S.S.R.
of Sciences of the USSR, Siberian Branch

OTHER ARTIFICIAL FIBRES   C.5.b)

Laboratoire de Biochimie et de Toxicologie   Canada
Pulmonaires, Département de Biologie,
Faculté des Sciences, Université de Sherbrooke

Laboratoire de caractérisation de l'amiante (LCA)   Canada
Faculté des Sciences, Université de Sherbrooke

Institut National de Recherche sur la Sécurité   France
(I.N.R.S.) (1)

Construction Materials Research Group, Civil   United Kingdom
Engineering Department, University of Surrey

ASBESTOS TAILINGS   C.6

Woodsreef Mines Limited (2)   Australia

Cassiar Mining Corporation   Canada

Centre spécialisé en technologie minérale   Canada

Laboratoire de caractérisation de l'amiante (LCA),   Canada
Faculté des Sciences, Université de Sherbrooke

Wayne State University, Occupational and Environmental   U.S.A.
Health

OTHER MATERIALS   C.7

Institute of Environmental Hygiene, University of   Austria
Vienna (1)(2)

LA 10 - CNRS & IOPG   France

Asahi Glass Co., Ltd., Research Laboratory   Japan

Service Cantonal d'Ecotoxicologie/Genève   Switzerland

Strangeways Research Laboratory   United Kingdom

ANY INDUSTRIAL APPLICATIONS   D.0

National Occupational Health & Safety Commission   Australia

Österreichische Staub (Silikose)-Bekämpfungsstelle   Austria

Micro & Trace Analysis Centre (MITAC),   Belgium
University of Antwerp (U.I.A.)

Service de Pneumologie, Hôpital Erasme   Belgium

Institut Armand-Frappier, Centre de recherche en   Canada
épidémiologie (1)

Laboratoire de caractérisation de l'amiante (LCA),   Canada
Faculté des Sciences, Université de Sherbrooke

Laboratoire de Biochimie et de Toxicologie   Canada
Pulmonaires, Département de Biologie,
Faculté des Sciences, Université de Sherbrooke

Health & Safety Executive, Epidemiology and   United Kingdom
Medical Statistics Unit

ASBESTOS CEMENT   <u>D.1</u>

Sydney University, Department of Mechanical Engineering   Australia

Institute of Environmental Hygiene, University of   Austria
Vienna (1)

Institut Armand-Frappier/Centre de recherche en   Canada
épidémiologie (2)

Ecole Supérieure de l'Energie et des Matériaux,   France
Université d'Orléans

Forschungs- und Entwicklungsabteilung, Hoechst AG,   Germany (FRG)
Werk Kelheim (1)

Laboratory of Biochemistry and Laboratory of Cellular   Israel
Immunology, Carmel Hospital

Asahi Glass Co., Ltd., Research Laboratory   Japan

Academy of Mining and Metallurgy, Interbranch   Poland
Institute of Building and Refractory Materials

Anjalena Publications Ltd. (2)   United Kingdom

Construction Materials Research Group, Civil   United Kingdom
Engineering Department, University of Surrey

MRC Environmental Epidemiology Unit   United Kingdom

Water Research Centre Engineering   United Kingdom

ASBESTOS TEXTILES (IN GENERAL)   <u>D.2</u>

Forschungs- und Entwicklungsabteilung, Hoechst AG,   Germany (FRG)
Werk Kelheim (2)

Hebrew University, Department of Medical Ecology   Israel

Anjalena Publications Ltd. (2)   United Kingdom

YARN AND CORDS   <u>D.2.a)</u>

Forschungs- und Entwicklungsabteilung, Hoechst AG,   Germany (FRG)
Werk Kelheim (2)

WOVEN GOODS   <u>D.2.b)</u>

Centre de Recherches sur la Physico-Chimie des   France
Surfaces Solides (CNRS) (2)

Forschungs- und Entwicklungsabteilung, Hoechst AG,   Germany (FRG)
Werk Kelheim (2)

INSULATION MATERIALS (IN GENERAL)   D.3

Academy of Mining and Metallurgy, Interbranch Institute of Building and Refractory Materials	Poland
Clinic of Occupational Medicine	Sweden
Anjalena Publications Ltd. (2)	United Kingdom
Respiratory Investigation Centre, Belfast City Hospital	United Kingdom
Kuakini Medical Center (1)(2)	U.S.A.
Mount Sinai School of Medicine (1)(2)(4)(5)(6)(7) (8)(9)(10)(11)(12)(13)(14)(15)	U.S.A.

FIRE-PROOF INSULATION   D.3.a)

Department of Community Health Sciences, Faculty of Medicine, The University of Calgary	Canada
Ontario Research Foundation	Canada
LA 10 - CNRS & IOPG	France
Asahi Glass Co., Ltd., Research Laboratory	Japan

HEAT INSULATION   D.3.b)

LA 10 - CNRS & IOPG	France

ACOUSTICAL INSULATION   D.3.c)

LA 10 - CNRS & IOPG	France

FRICTION MATERIALS (IN GENERAL)   D.4

Beral Bremsbelag GmbH   Germany (FRG)

Hebrew University, Department of Medical Ecology   Israel

Anjalena Publications Ltd. (2)   United Kingdom

CLUTCHES   D.4.b)

Forschungs- und Entwicklungsabteilung, Hoechst AG,   Germany (FRG)
Werk Kelheim (2)

SEALING MATERIALS (IN GENERAL)   D.5

Anjalena Publications Ltd. (2)   United Kingdom

SEALING MATERIALS (OTHER THAN JOINTS AND CAULKING)   D.5.c)

Mine Safety and Health Administration (MSHA),   U.S.A.
U.S. Department of Labor (1)

FILLERS   <u>D.6</u>

Centre de Recherches sur la Physico-Chimie des   France
Surfaces Solides (CNRS) (1)

Anjalena Publications Ltd. (2)   United Kingdom

PAPER PRODUCTS   <u>D.7</u>

Anjalena Publications Ltd. (2)   United Kingdom

ASPHALT MIX   <u>D.8</u>

Anjalena Publications Ltd. (2)   United Kingdom

FILTERS AND DIAPHRAGMS   <u>D.9</u>

Département de Chimie, Université de Sherbrooke   Canada

Centre de Recherches sur la Physico-Chimie des   France
Surfaces Solides (CNRS) (2)

Forschungs- und Entwicklungsabteilung, Hoechst AG,   Germany (FRG)
Werk Kelheim (2)

Anjalena Publications Ltd. (2)   United Kingdom

REFRACTORY PRODUCTS   <u>D.10</u>

Anjalena Publications Ltd. (2)   United Kingdom

GASKETS AND PACKINGS (IN GENERAL)   D.11

Anjalena Publications Ltd. (2)   United Kingdom

GASKETS   D.11.a)

Mine Safety and Health Administration (MSHA),   U.S.A.
U.S. Department of Labor (2)

PACKINGS   D.11.b)

Martin Merkel GmbH & Co, KG   Germany (FRG)

FLOOR TILES/PLASTICS   D.12

Institute of Environmental Hygiene,   Austria
University of Vienna (2)

Laboratoire d'Etude des Particules Inhalées (LEPI)(3)   France

Anjalena Publications Ltd. (2)   United Kingdom

OTHER APPLICATIONS   D.13

Laboratory of Catalysis, Department of Chemistry, Canada
Concordia University

Forschungs- und Entwicklungsabteilung, Hoechst AG, Germany (FRG)
Werk Kelheim (2)

ANY SUBSTITUTES   E.0

National Occupational Health & Safety Commission   Australia

Sydney University, Department of Mechanical Engineering   Australia

Micro & Trace Analysis Centre (MITAC),   Belgium
University of Antwerp (U.I.A)

Laboratoire de caractérisation de l'amiante (LCA),   Canada
Faculté des Sciences, Université de Sherbrooke

Programme de recherche sur l'amiante,   Canada
Université de Sherbrooke (2)(4)

Beral Bremsbelag GmbH   Germany (FRG)

ARAMID FIBRES   E.2

Sydney University, Department of Mechanical Engineering   Australia

Laboratoire de caractérisation de l'amiante (LCA),   Canada
Faculté des Sciences, Université de Sherbrooke

Anjalena Publications Ltd. (2)   United Kingdom

BEATER-SATURATION MATERIALS   E.3

Anjalena Publications Ltd. (2)   United Kingdom

CARBON   E.4

Martin Merkel GmbH & Co, KG   Germany (FRG)

Anjalena Publications Ltd. (2)   United Kingdom

CERAMICS   E.5

Laboratoire de caractérisation de l'amiante (LCA),   Canada
Faculté des Sciences, Université de Sherbrooke

Laboratoire de Biochimie et de Toxicologie   Canada
Pulmonaires, Département de Biologie,
Faculté des Sciences, Université de Sherbrooke

Anjalena Publications Ltd. (2)   United Kingdom

CLAYS   E.6

Laboratoire de caractérisation de l'amiante (LCA),   Canada
Faculté des Sciences, Université de Sherbrooke

Laboratoire de Biochimie et de Toxicologie   Canada
Pulmonaires, Département de Biologie,
Faculté des Sciences, Université de Sherbrooke

I.N.S.E.R.M. Unité 139 (1)(3)   France

Anjalena Publications Ltd. (2)   United Kingdom

GLASS FIBRES   E.11

Laboratoire de caractérisation de l'amiante (LCA)   Canada
Faculté des Sciences, Université de Sherbrooke

Laboratoire de Biochimie et de Toxicologie   Canada
Pulmonaires, Département de Biologie,
Faculté des Sciences, Université de Sherbrooke

LA 10 - CNRS & IOPG   France

Asahi Glass Co., Ltd., Research Laboratory   Japan

Clinic of Occupational Medicine   Sweden

Anjalena Publications Ltd. (2)   United Kingdom

GRAPHITE   E.12

Martin Merkel GmbH & Co., KG   Germany (FRG)

Anjalena Publications Ltd. (2)   United Kingdom

GRAPHITE TFE COMPOSITE   E.13

Martin Merkel GmbH & Co., KG   Germany (FRG)

Anjalena Publications Ltd. (2)   United Kingdom

METAL   E.15

Anjalena Publications Ltd. (2)   United Kingdom

MICA   E.16

Anjalena Publications Ltd. (2)   United Kingdom

NYLON   E.18

Anjalena Publications Ltd. (2)   United Kingdom

ORGANIC FELTS   E.19

Anjalena Publications Ltd. (2)   United Kingdom

PLASTICS   E.20

Anjalena Publications Ltd. (2)   United Kingdom

PLATEY TALC   E.21

Anjalena Publications Ltd. (2)   United Kingdom

PMF   E.22

Laboratoire de caractérisation de l'amiante (LCA),   Canada
Faculté des Sciences, Université de Sherbrooke

Anjalena Publications Ltd. (2)   United Kingdom

POLYETHYLENE   E.23

Anjalena Publications Ltd. (2)   United Kingdom

POLYMERS   E.24

Sydney University, Department of Mechanical Engineering   Australia

Anjalena Publications Ltd. (2)   United Kingdom

POLYURETHANE   E.25

Anjalena Publications Ltd. (2)   United Kingdom

PVC   E.26

Anjalena Publications Ltd. (2)   United Kingdom

REINFORCED CONCRETE   E.27

Anjalena Publications Ltd. (2)   United Kingdom

RUBBER   E.28

Anjalena Publications Ltd. (2)   United Kingdom

SEMIMETALLICS   E.29

Anjalena Publications Ltd. (2)   United Kingdom

STAPLE GLASS   E.30

Anjalena Publications Ltd. (2)   United Kingdom

STEEL FIBRES   E.31

Anjalena Publications Ltd. (2)   United Kingdom

TFE   E.32

Anjalena Publications Ltd. (2)   United Kingdom

VEGETABLE FIBRE SHEET   E.33

Laboratoire de caractérisation de l'amiante (LCA),   Canada
Faculté des Sciences, Université de Sherbrooke

Anjalena Publications Ltd. (2)   United Kingdom

WOLLASTONITE   E.35

Laboratoire de caractérisation de l'amiante (LCA), Faculté des Sciences, Université de Sherbrooke	Canada
Laboratoire de Biochimie et de Toxicologie Pulmonaires, Département de Biologie, Faculté des Sciences, Université de Sherbrooke	Canada
Institute of Occupational Health (2)	Finland
Anjalena Publications Ltd. (2)	United Kingdom

OTHER SUBSTITUTES   E.37

Forschungs- und Entwicklungsabteilung, Hoechst AG, Werk Kelheim (1)(2)	Germany (FRG)
Asahi Glass Co., Ltd., Research Laboratory	Japan
Clinic of Occupational Medicine	Sweden
Construction Materials, Research Group, Civil Engineering Department, University of Surrey	United Kingdom
Mine Safety and Health Administration (MSHA), U.S. Department of Labour (1)(2)	U.S.A.

Annex

CONTINUATION OF CERTAIN DESCRIPTIONS

p. 29   Publication(s) on the project:

2. Siemiatycki J. An epidemiologic approach to discovering occupational carcinogens by obtaining better information on occupational exposures. In Recent Advances in Occupational Health. Ed. M.J. Harrington, Vol. 2, pp 143-157, 1984.
3. Gérin M., Siemiatycki J, Kemper H, Bégin D. Obtaining occupation exposure histories in epidemiologic case-control studies. J. Occup. Med. 27:420-426, 1985.

p. 32   Publication(s) on the project:

Ménard, H., L. Noël, J. Khorami, J.L. Jouve and J. Dunnigan. The adsorption of polyaromatic hydrocarbons on natural and chemical modified asbestos fibers. Environ. Res. (Accepted for publication, 1985).
Various technical reports on natural or man-made fibers (biological activity and physico-chemical characterization).

p. 46   Research protocol:

The individual neurological findings did not correlate with the results of pulmonary function tests or with most of the immunological data. The tumor markers (CEA, ferritin, b2-microglobulin) also did not correlate with the individual neurological findings even though increased levels of rheumatoid factor and Wa-Ro titers were more frequent among patients with dysfunction in both the peripheral and the central nervous systems. Also increased levels of the three tumor markers were found for the asbestosis patients more often than for the referents. During the follow-up 13 asbestosis patients have developed cancer.
The follow-up will be continued until the end of 1986.

Publication(s) on the project:

Huuskonen M., Räsänen J., Juntunen J. & Partanen T.: Immunological aspects of asbestosis: Patients' neurological signs and asbestosis progression. Am.j.industr.med. 5 (1984) 461-469.

p. 47   Research protocol:

In in vitro tests, wollastonite had very low hemolytic activity to human red blood cells when compared to that of DQ 12 silica and chrysotile asbestos. Wollastonite was slightly cytotoxic to rat alveolar macrophages.

Publication(s) on the project:

Huuskonen M.S., Tossavainen A., Koskinen H., Zitting A., Korhonen O., Nickels J., Korhonen K & Vaaranen V.: Wollastonite exposure and lung fibrosis, Environm.res. 30 (1983) 291-304.
Huuskonen M.S., Tossavainen A, Koskinen H, Zitting A, Korhonen O, Nickels J, Korhonen K & Vaaranen V.: Respiratory morbidity of quarry workers exposed to wollastonite In: Occupational lung disease. Ed. by B.E. Gee, W.K.C. Morgan and S.M. Brooks, Raven Press, New York 1984, p. 245.

p. 84 Cooperation:

C. Bianchi Unità Sanitaria Locale n. 2 - Goriziana, Presidio Ospedaliero di Monfalcone.

C. Ciallella Istituto di Medicina Legale, Università "La Sapienza", Roma;

M.A. Dina Istituto di Anatomia ed Istologia Patologica, Università Cattolica del Sacro Cuore, Roma;

A. Donna Servizio di Anatomia e Istologia Patologica, Ospedale Generale Provinciale di Alessandria;

C. Maltoni Istituto di Oncologia "F. Addarii", Università di Bologna;

A.M. Mancini Panel Nazionale dei Mesoteliomi. Istituto di Anatomia e Istologia Patologica, Università di Torino;

F. Mollo Istituto di Anatomia ed Istologia Patologica, Università di Torino;

F. Nardi Dipartimento di Biopatologia, Università "La Sapienza", Roma.

ADDRESSES OF PARTICIPATING RESEARCH INSTITUTIONS

Page

Name :	ACADEMY OF MINING AND METALLURGY, INTERBRANCH INSTITUTE OF BUILDING AND REFRACTORY MATERIALS	90
Address :	Al. Mickiewicza 30, 30-059 Krakow, Poland	
Phone :	33-91-00 w. 23-35 Telex : 032203 pl	

Name :	ANJALENA PUBLICATIONS LTD.	100
Address :	65 Wellesley Drive, Crowthorne, Berkshire RG11 6AL, United Kingdom	
Phone :	0344 771743 Telex :	

Name :	ASAHI GLASS CO., LTD., RESEARCH LABORATORY	87
Address :	Hazawa-cho Kanagawa-ku, Yokohama 221, Japan	
Phone :	045-381-1441 Telex : J 24616 ASAGLAS	

Name :	ASBESTOS RESEARCH GROUP, QUEEN ELIZABETH II MEDICAL CENTRE	9
Address :	Nedlands, Western Australia, Australia	
Phone :	3893333 Telex :	

Name :	BERAL BREMSBELAG GmbH	69
Address :	Postfach 1160, D-5277 Marienheide, Germany (FRG)	
Phone :	02264/89-1 Telex : 884 192 beral d	

Name :	BERUFSGENOSSENSCHAFTLICHES INSTITUT FÜR ARBEITSSICHERHEIT	70
Address :	Lindenstrasse 80, Postfach 2043, D-5205 Sankt Augustin 2, Germany (FRG)	
Phone :	(02241) 231-02 Telex : 889 460 bia d	

Name :	CASSIAR MINING CORPORATION	20
Address :	Cassiar, British Columbia, VOC 1EO, Canada	
Phone :	(604) 778 7435 Telex : 036 88533	

Name :	CENTRE DE RECHERCHES SUR LA PHYSICO-CHIMIE DES SURFACES SOLIDES (CNRS)	50
Address :	24, avenue du Président Kennedy, F-68200 Mulhouse, France	
Phone :	89 42 01 55 Telex :	

Name :	CENTRE SPECIALISE EN TECHNOLOGIE MINERALE	21
Address :	671, boulevard Smith sud, Thetford-Mines, Québec (Qué.) G6G 1N1, Canada	
Phone :	(418)338-8591 Telex :	

Page

Name :	CENTRO STUDI E RICERCHE SUGLI EFFETTI BIOLOGICI DELLE POLVERI INALATE, ISTITUTO DI MEDICINA DEL LAVORO, UNIVERSITA DI MILANO	79
Address :	Via S. Barnaba 8, I-20122, Milano, Italy	
Phone :	(2) 545.7324	Telex :

Name :	CENTRO DI STUDIO PER I PROBLEMI MINERARI c/o DIPARTIMENTO GEORISORSE E TERRITORIO-POLITECNICO	83
Address :	C.so Duca degli Abruzzi 24, I-10129 Torino, Italy	
Phone :	(011) 556.7618	Telex :

Name :	CLINIC OF OCCUPATIONAL MEDICINE	92
Address :	S-104 01 Stockholm, Sweden	
Phone :	08 - 736 3056	Telex :16660 (KARO)

Name :	CLINIQUE TOXICOLOGIQUE, HOPITAL FERNAND-WIDAL	52
Address :	200, rue du Faubourg Saint-Denis, F-75475 Paris, France	
Phone :	42.80.62.33	Telex :

Name :	CONSTRUCTION MATERIALS RESEARCH GROUP, CIVIL ENGINEERING DEPARTMENT, UNIVERSITY OF SURREY	102
Address :	Guildford, United Kingdom	
Phone :	0483 859331	Telex : 859331

Name :	DEPARTEMENT DE CHIMIE, UNIVERSITE LAVAL	25
Address :	Cité Universitaire, Québec, Qué. G1 K7P4, Canada	
Phone :	418-6562131	Telex :

Name :	DEPARTEMENT DE CHIMIE, UNIVERSITE DE SHERBROOKE	26
Address :	Cité Universitaire, Boulevard de L'Université, Sherbrooke, Qué. J1K 2R1, Canada	
Phone :	819-8217000	Telex :

Name :	DEPARTMENT OF COMMUNITY HEALTH SCIENCES, FACULTY OF MEDICINE, THE UNIVERSITY OF CALGARY	22
Address :	3330 Hospital Drive N.W., Calgary, Alberta T2N 4N1, Canada	
Phone :	(403) 220-4297	Telex :

Name :	DEPARTMENT OF COMMUNITY MEDICINE, TOYAMA MEDICAL AND PHARMACEUTICAL UNIVERSITY	88
Address :	2630 Sugitani, Toyama City, Toyama Prefecture, 930-01, Japan	
Phone :	0764-34-2281 (ext. 2370 - 2373)	Telex :

199

Page

Name :   DEPARTMENT OF HYGIENE, KAWASAKI MEDICAL SCHOOL   89

Address : Matsushima 577, Kurashiki, 701-01, Japan
Phone :   0864-62-1111   Telex :

Name :   DEPARTMENT OF MINERAL AND ENERGY AFFAIRS,   91
          AIR QUALITY RESEARCH
Address : P.O. Box 1132, Johannesburg 2000, South Africa
Phone :   011-4025415/9   Telex :

Name :   DEPARTMENT OF ONCOLOGY, UNIVERSITY HOSPITAL UMEA   93

Address : S-901 87 Umeå, Sweden
Phone :   Telex :

Name :   DEPARTMENT OF PHYSICS, DALHOUSIE UNIVERSITY   23

Address : Halifax, N.S., Canada B3H 3J5
Phone :   (902) 424-7062/2337   Telex :

Name :   DEPARTMENT OF SOIL SCIENCE, UNIVERSITY OF BRITISH COLUMBIA   24

Address : Vancouver, B.C. V6T 2A2, Canada
Phone :   604-228-4898   Telex :

Name :   EAST BAY MUNICIPAL UTILITY DISTRICT   119
Address : 2127 Adeline, P.O. Box 24055, Oakland, Ca 94623, U.S.A.
Phone :   (415) 891-0683   Telex :

Name :   ECOLE SUPERIEURE DE L'ENERGIE ET DES MATERIAUX   53
          UNIVERSITE D'ORLEANS
Address : BP 6749, F-45067 Orléans, Cedex 2, France
Phone :   38.63.22.16   Telex :

Name :   ENVIRONMENTAL POLLUTION RESEARCH CENTRE,   76
Address : Department of Respiratory Medicine, Cardiovascular &
          Thoracic Centre, Seth G.S. Medical College &
          K.E.M. Hospital, Bombay 400 012, India
Phone :   41 32 296 & 41 36 051 (Ext. 5)   Telex :

Name :   FACULTE DE MEDECINE, DEPARTEMENT DE PNEUMOLOGIE,   27
          UNIVERSITE DE SHERBROOKE
Address : Sherbrooke, Québec, Canada J1K 2R1
Phone :   (819) 565-2196   Telex : 05-836149

Page

Name : ISTITUTO SUPERIORE DI SANITA,   84
       LABORATORIO DI ULTRASTRUTTURE

Address : Viale Regina Elena 299, I-00161 Roma, Italy

Phone : (06) 4990 ext. 228   Telex : 610071 ISTSAN I

Name : ITALIAN RAILWAY HEALTH SERVICE,   85
       SERVIZIO SANITARIO FERROVIE DELLO STATO

Address : P.zza della Croce Rossa 1, I-00100 Roma, Italy

Phone : 06/856493   Telex :

Name : JOHNS HOPKINS SCHOOL OF HYGIENE AND PUBLIC HEALTH   123

Address : 615 N. Wolfi st., Baltimore, MD 21205, U.S.A.

Phone : 301-955-3906   Telex :

Name : KUAKINI MEDICAL CENTER   124

Address : 347 N. Kuakini Street, Honolulu, HI 96817, U.S.A.

Phone : (808) 547-9532   Telex :

Name : LA 10 - CNRS & IOPG   65
       LA = Laboratoire Associé n° 10 du CNRS
       IOPG = Institut et Observatoire de Physique du Globe

Address : 5 rue Kessler, F-63038 Clermont-Ferrand Cedex, France

Phone : 73 93 35 71   Telex : 392-450

Name : LABORATOIRE DE BIOCHIMIE ET DE TOXICOLOGIE PULMONAIRES,   31
       DEPARTEMENT DE BIOLOGIE, FACULTE DES SCIENCES,
       UNIVERSITE DE SHERBROOKE

Address : Sherbrooke, QC, Canada J1K 2R1

Phone : (819) 821-7635 (ext. 3072)   Telex : 05-836149

Name : LABORATOIRE DE CARACTERISATION DE L'AMIANTE,   32
       FACULTE DES SCIENCES, UNIVERSITE DE SHERBROOKE

Address : 2500, boul. Université, Sherbrooke, Québec, Canada J1K 2R1
Phone : (819) 821-7635 (ext. 3072)   Telex : 05-836149

Name : LABORATOIRE D'ETUDE DES PARTICULES INHALEES (LEPI)   66

Address : 44, rue Charles Moureu, F-75013 Paris, France

Phone : (1) 45.84.12.09 (ext. 152)   Telex :

Name : LABORATORY OF BIOCHEMISTRY AND LABORATORY OF CELLULAR   78
       IMMUNOLOGY, CARMEL HOSPITAL

Address : Rotshild bld. 35-37, Haifa, Israel

Phone : 532131 & 250211   Telex :

203

Page

Name :	LABORATORY OF CATALYSIS, DEPARTMENT OF CHEMISTRY, CONCORDIA UNIVERSITY	33

Address : 1455 de Maisonneuve Blvd., West Montreal, QC, Canada H3G 1M8

Phone : (514) 848-3343   Telex :

Name : MARTIN MERKEL GmbH & Co. KG   74

Address : Sanitasstrasse 17-21, P.O. Box 93 02 80, D-2102 Hamburg 93, Germany (FRG)

Phone : (040) 75 11-0   Telex : 2 16 35 22 mer d

Name : MICRO & TRACE ANALYSIS CENTRE (MITAC), UNIVERSITY OF ANTWERP (U.I.A)   18

Address : Universiteitsplein 1, B-2610 Wilrijk, Belgium

Phone : 03/828.25.28 (ext. 120 or 213)   Telex : 33646

Name : MINE SAFETY AND HEALTH ADMINISTRATION, U.S. DEPT. OF LABOR   126

Address : Pittsburgh Health Technology Center, 4800 Forbes Avenue Pittsburgh, Pa. 15213, U.S.A.

Phone : (412) 621-4500 X252   Telex :

Name : MOUNT SINAI SCHOOL OF MEDICINE   128

Address : 1 Gustave L. Levy Place, New York, NY 10029, U.S.A.

Phone : (212) 650-7809   Telex : Mt. Sinai Hospital 968890

Name : MRC ENVIRONMENTAL EPIDEMIOLOGY UNIT   114

Address : Southampton General Hospital, Southampton, S09 4XY, U.K.

Phone : 0703 777624   Telex : 47661 SOTONU G

Name : NATIONAL BOARD OF OCCUPATIONAL SAFETY AND HEALTH   94

Address : S-171 84 Solna, Sweden

Phone : 08-7309000   Telex : 158 16 ARBSKY S

Name : NATIONAL INSTITUTE OF ENVIRONMENTAL HEALTH SCIENCES, PULMONARY PATHOLOGY LABORATORY   143

Address : P.O. Box 12233, Res. Tri. Pk., NC 27709

Phone : (919) 541-3243 or 4410   Telex :

Name : NATIONAL INSTITUTE FOR OCCUPATIONAL SAFETY AND HEALTH (NIOSH)   146

Address : 4676 Columbia Pkwy, Cincinnati, OH 45226, U.S.A.

Phone : 513-841-4266   Telex :

Page

Name :	SCHOOL OF OCCUPATIONAL HEALTH, McGILL UNIVERSITY	42
Address :	1130 Pine West Avenue, Montréal, Québec, Canada H3A 1A3	
Phone :		Telex :

Name :	SERVICE CANTONAL D'ECOTOXICOLOGIE, GENEVE	99
Address :	23, avenue Ste Clotilde, CH-1211 Genève, Switzerland	
Phone :	022/28.75.11	Telex :

Name :	SERVICE DE PNEUMOLOGIE, HOPITAL ERASME	19
Address :	Route de Lennik 808, B-1070 Brussels, Belgium	
Phone :	2/568.39.43	Telex :

Name :	SERVIZIO DI ANATOMIA E ISTOLOGIA PATOLOGICA OPEDALE DI MONFALCONE	86
Address :	via Galvani, I-34074 Monfalcone, Italia	
Phone :	0481-470101	Telex :

Name :	STRANGEWAYS RESEARCH LABORATORY	117
Address :	Worts Causeway, Cambridge CB1 4RN, United Kingdom	
Phone :	(0223) 243231	Telex :

Name :	SYDNEY UNIVERSITY, DEPARTMENT OF MECHANICAL ENGINEERING	12
Address :	Sydney, Australia	
Phone :	02-692-2285	Telex : AA26169

Name :	UNIVERSITY OF UPPSALA, DEPARTMENT OF LUNG MEDICINE	97
Address :	University Hospital, S-751 85 Uppsala, Sweden	
Phone :	018-166000	Telex :

Name :	WATER RESEARCH CENTRE ENGINEERING	118
Address :	P.O. Box 85, Frankland Road, Blagrove, Swindon, Wilts. SN5 8YR, United Kingdom	
Phone :	0793 488301	Telex : 449541

Name :	WAYNE STATE UNIVERSITY, OCCUPATIONAL AND ENVIRONMENTAL HEALTH	147
Address :	Detroit, Michigan 48202, U.S.A.	
Phone :	313-577-1210	Telex :

Name :	WOODSREEF MINES LIMITED	13
Address :	13, Flynn Avenue, Barraba NSW 2347, Australia	
Phone :	(067) 821401	Telex : 163855

LIST OF SCIENTISTS RESPONSIBLE FOR THE STUDIES

AND OF OTHER SCIENTISTS INVOLVED

SUBJECT INDEX

221

	Classification code	This Directory (page)
Valorization		
see: mineralurgy		
Vegetable fibre sheet	E.33	189
Waste management		
in general	B.10	168
disposal	B.10.a)	168
Wollastonite	E.35	190
Woven goods	D.2.b)	178
Yarn	D.2.a)	178